HIGHLY SENSITIVE PERSONS WITH HIGH-LEVELS OF EMPATHY

The Survival Guide

Life Strategies For Eliminating Negative Emotions, Managing Relationships, And Thriving As A Highly Sensitive Person

By

Baylee Martin

© **Copyright 2020 by Baylee Martin. All rights reserved.**

This document aims to provide accurate and reliable information about the topic and issue covered. This publication is sold with the understanding that the publisher is not required to render any qualified services such as accounting, officially permitted, or otherwise. A practiced individual in the profession should be sought, if professional advice is required.

This information is from the Declaration of Principles which was accepted and approved by both the Committee of the American Bar Association and the Committee of Publishers and Associations.

In no way is it legal to reproduce, duplicate, or transmit any part of this document by either electronic means or printed format. Recording of this publication is strictly prohibited, and any storage of this document is also prohibited unless written permission from the publisher is granted. All rights reserved.

The information provided herein is stated to be truthful and consistent, in that any liability, in terms of inattention or otherwise, by any usage or abuse of any policies, processes, or directions contained within is the solitary and only responsibility of the recipient reader. Under no circumstances will any legal responsibility or blame be held against the publisher for any reparation, damages, or monetary loss due to the information herein, either directly or indirectly.

Respective authors own all copyrights not held by the publisher.

The information herein is offered for informational purposes solely and is universally understood. The presentation of the information is without a contract or any type of guarantee assurance.

The trademarks that are being used are done so without any consent from the present owner, and the publication of the trademark is without permission or backing by the trademark owner. All trademarks and brands within this book are for clarifying purposes only and are owned by the owners themselves, not affiliated with this document.

TABLE OF CONTENTS

Introduction 1
Chapter 1: Empaths, Emotions, And Health 18
Chapter 2: The Difference Between Introverts, Empaths, And Highly Sensitive People 28
Chapter 3: Empaths, Love And Sex 52
Chapter 4: Protecting Yourself From Narcissists And Other Energy Vampires 69
Chapter 5: Empaths, Parenting, And Raising Sensitive Children. 81
Chapter 6: Empaths And Work 105
Chapter 7: Empaths, Intuition, And Extraordinary Perceptions . 114
Chapter 8: The Gift Of Being An Empath 119
Chapter 9: Empaths And Addiction: From Alcohol To Overeating 129
Conclusion 138

INTRODUCTION

Are you affected by the feelings of everyone around you? Do individuals describe you as compassionate? Maybe you consistently feel the emotions and physical symptoms of others, as if they are your own. If this rings true, you might be an "empath."

Just 1 to 2 percent of the population experiences this kind of affectability, that is, being able to feel and assimilate the emotions that encompass them. These people are more likely to view the world primarily with their feelings and instincts, opposed to putting an excessive amount of rationale behind each decision. While this characteristic can be a wellspring of individual quality, it is essential to realize how to overcome the everyday challenges of being an empath.

What is an Empath?

While there is a large body of research explaining the feeling of empathy, there are few studies that concentrate on the principles of being an empath. What is known is that, as indicated by studies, empaths likely have hyper-responsive mirror neurons — the group of brain cells answerable for activating feelings like empathy. For example, it is feasible for somebody to feel particularly sensitive to electromagnetic fields created by an individual's brain and heart and intuit the feelings of the people around them. So, if there is an energized crowd or a group of individuals in mourning, the energy generated by them can be profoundly felt inside an empath's body.

For individuals who are introverted empaths, they might be increasingly sensitive to the brain chemical responsible for feeling joy — dopamine. In cases where an excess of excitement occurs, an empath can feel overpowered. After some time, empaths begin to avoid external stimulation and learn to live without it to feel happy. Regardless of whether an individual is contemplative, some common side effects of hypersensitivity include weariness, over-burden, depression and nervousness. When these emotions emerge, it's useful to have some space in your home to withdraw or even a favorite spot outside in the open air, any place you can energize in.

At any point when an empath is overwhelmed with stressful emotions, anxiety, panic attacks, depression and weakness can occur. They may even show physical side effects, for example, an increased pulse and headache. This is thought to occur because they disguise the feelings and pain of others, and they don't have the awareness to distinguish them from their own. Helping empaths to deal with these experiences is essential; they need to be able to separate their feelings and emotions, as best as they can, from those around them.

Utilizing Your Emotions as a Strength

For around 15 to 20 percent of the population, referred to as "profoundly delicate," they feel more deeply and vigorously than everyone around them. Their minds handle and consider information in an incredible, nuanced manner. While this behavior can be viewed as being excessively sensitive, mindful or excessive, they can likewise be seen as positive — being outstandingly insightful, intuitive and hyper-observant. Empaths should try to figure out how to focus on and channel those unforgettable emotions.

To prepare for terrible or overpowering feelings, empaths can turn to a variety of coping strategies to make everyday encounters progressively better. It helps to adopt a methodological strategy for time management, and set strict boundaries with individuals who you know channel your energy. Also, knowing when meditation and stillness are helpful is important for maintaining calm. The key is to discover which approaches help you and then deliberately use them to counteract elevated emotions as they emerge.

While more research is required to better understand the science behind empaths and the reasons why only a few people process emotions more acutely than others, there are ways to check whether you are experiencing empath-like tendencies. By understanding the signs of being an empath and what triggers symptoms, a person can learn to transform them into beneficial qualities, while at the same time coping with any adverse effects. For example, it is essential to identify your emotional needs and convey them to the people around you.

Signs That You're an Empath

An empath is somebody who is exceptionally aware of the emotions of everyone around them, to the point of feeling those emotions themselves. Empaths see the world uniquely; they're distinctly aware of the emotional state of others, what causes them pain and their emotional needs.

However, it's not that Empaths simply feel other people's emotions. The Empath's Survival Guide will explain, how empaths feel physical pain as well, and how they can regularly detect — the goals of other people and where their emotional states are coming from. Empaths appear to pick up on a lot about the people around them.

Numerous highly sensitive people can also be empaths — yet there are also differences between empaths and HSPs. Having a high level of empathy is only one of the four attributes that makes somebody an HSP. They are more responsive to a number of interventions, notwithstanding emotional ones. What is clear is that most empaths are highly sensitive, yet not all susceptible individuals are fundamentally empaths.

13 Signs of an Empath

1. You take on other people's emotions as your own

This is the ultimate sign, the number one characteristic of an empath. Regardless of what another person close to you is feeling, regardless of whether they figure they aren't indicating it, if you are an empath, you will probably pick up on it right away. Yet, more than that: you may feel the emotion as though it were your own, basically "absorbing" it or soaking it up.

How precisely this works is a subject of discussion. In any case, we do know that individuals who have significant levels of empathy likewise have extremely active mirror neurons — the part of the brain that read emotional prompts from others and makes sense of what they may be thinking or feeling. In case you're an empath, you catch onto subtle changes in expression, non-verbal communication, or manners of speaking that others miss — and then promptly embody what the individual is feeling.

However, those same active mirror neurons cause you to mainly live through the feeling of others as though they were your own. That can be a powerful gift, albeit exhausting and overpowering at times.

2. In some cases, you experience unexpected, overpowering emotions when you're in public

It's not simply in one-on-one discussions where you sense the emotions of others. It can all of a sudden occur whenever others are present.

As an empath, going into public spaces becomes a challenge since you may unexpectedly be overwhelmed with an emotion that came out of nowhere. Of course, an empath knows that emotion came from another person in the area, but they don't know from whom.

3. The "vibe" of a room matters to you — a lot

There is no doubt that empaths are susceptible to the "feel" or dynamics of their environment. But when surrounded by harmony and stillness, they prosper, because they take on those characteristics inside themselves. For example, moments of magnificence can be transformative for empaths, whether it's a peaceful garden, a beautiful room or the corridors of an exhibition hall. In the same way, if conditions become disordered or discouraging, the energy of the empath will rapidly drain out of them.

4. You comprehend where individuals are coming from

Empath experts describe this as the core quality of an empath — considerably more so than absorbing the feelings of others. Empaths can learn not to retain feelings to such an extent, and some empaths once in a while don't "absorb" them at all. However, all empaths can instinctively detect what somebody is attempting to communicate, even when they're having a hard time expressing themselves.

In a general sense, empathy is about comprehending and

connecting with others. What's more, it's about sensing where individuals are coming from.

5. Friends go to you for advice

Empaths are regularly searched out by their companions for counsel, support, and encouragement because they have the ability to understand others so deeply. It helps that empaths, in general, are good listeners who accept others and are trustworthy. They are consistent because they react from the heart and mean what they say.

If this sounds like you, you probably realize that being an empath is hard. Rarely, people acknowledge the amount of personal energy it takes for you to be a good listener or to give advice. A few people even underestimate it.

6. Tragic and violent events on TV can weaken you

For the empath, it doesn't make a difference if a horrendous event is happening to you directly or indirectly because, despite everything, you feel it through your very existence. You may appear to "live through" the pain or loss of the occasion as if it is happening to you directly. It doesn't matter if you are miles away from it — or even if it's a fictional event in a show. This response can be overpowering at times.

Empaths, as HSPs, should avoid watching violent or human tragedy movies, even if it is a popular movie.

7. You can't contain your love for pets, animals, or

children

Indeed, everybody realizes that children are delightful little miracles, and cats and dogs are adorable — yet for you, those feelings appear to be a lot more grounded. You will be unable to help yourself from gushing over somebody's lovely child or promptly hunching down to adore a little dog. A few people may say your response is "over the top," however, you will find yourself wondering why everyone doesn't have the same capacity.

Many agree that this is one of the many advantages of being an empath. All your feelings, including positive ones, are exaggerated.

8. You may feel an individual's physical diseases as well — not merely their emotions

At the point when somebody becomes ill or is harmed, you feel the illness to such an extreme that you experience it as though it's your own. This doesn't simply mean you're feeling sympathy or worried; instead, you experience genuine physical sensations like pain, muscle tightness, or irritation in similar zones of the body. Sometimes it is because your empathic brain is reflecting what the other individual hasn't realized yet. In addition, you anticipate their experience in your own body.

Furthermore, it can feel awkward — in any event and weakening. It's not a "blessing" that empaths enjoy. But on the other hand, it's at the base of why empaths are such exceptional caregivers. Without this capacity, they wouldn't have the ability to associate with somebody who is in pain or get them precisely what they need to feel better.

It's unsurprising that empaths are attracted to jobs like nursing, doctor, elderly provider, or healer. If you can sympathize with everybody's pain, it would be unlikely you wouldn't want to do something about it.

9. You can become overpowered in intimate relationships

Relationships can be challenging for everybody. However, envision how much greater those difficulties are if you can detect every temperament, disturbance, or lie from your partner. What's more, positive feelings can likewise get overpowering — as though the relationship may "engulf" you. Sound familiar?

There is more to being an empath partner. When you live together in a shared environment, it can become an obstacle. The person's "energy" you share space with is continuously present for an empath, and sometimes feel like a distraction. Empaths see their homes as a haven where they can escape from the steady interest in their emotional senses, and a partner can change that.

Consequently, some empaths decide to stay single, while others figure out how to adjust — maybe by having a room that is their private space or (critical) looking for a partner who respects their boundaries.

10. You're a walking lie detector

Without a doubt, there most likely have been times when somebody effectively tricked you and for some reason you ignore your gut instinct that was warning you from the beginning. The thing about an empath's capacity to process even the smallest expressive gestures implies that it's practically unimaginable for somebody to shroud their actual aims. Regardless of whether you fail to know precisely what an individual truly needs, you know whether they're not completely honest— or they are being dishonest about one particular thing.

11. You can't comprehend why any leader wouldn't put their teams first

There are a lot of managers and group coordinators who just don't focus on their group's needs. If you're an empath, this isn't simply discourteous or irritating — it's a failure of leadership.

This is the case because empaths make astounding leaders themselves. When they are in a leadership role, they consistently tune into their group and join everyone around shared objectives. Empaths, in general, are keen and mindful and ensure each team member feels heard. The outcome isn't only a more joyful group of individuals; it results in the creation of better choices by assimilating all the information.

12. You have a calming effect on others — and the ability to recuperate them

It's true. When individuals seek out empaths for advice, they feel comfortable in an empath's presence. It's no accident that individuals inadvertently seek their most empathic companions during troublesome events.

This is something you, as an empath, can develop and use to help individuals to recuperate. It is done with the desire to help them work past genuine psychological baggage and beat unfortunate circumstances. However, you can't do as such if you conceal your affectability and empathy — you need to grasp this blessing if you truly want to have any kind of effect.

13. You can't see somebody in pain without needing to help

Would you be able to stroll past somebody who's out of luck, without thinking about how you could support them? Do you struggle to turn off your anxiety for others because "there's work to do"? If your response is no — and not, only if you have the time, or not in a hurry — then, there's a good possibility you're an empath.

Furthermore, this is the reason empaths are such a significant piece of the astounding kaleidoscope of humankind. For an empath, individuals are the most beautiful things on their radar, and it's inconceivable not to see — and react to — the needs of others. That is actually where an empath's healing ability originates from, and it's something we could utilize in greater amounts in our lives.

What is Empathy?

Sympathy is, at its least complex, the consciousness of the feelings and emotions of others. It is a key component of Emotional Intelligence, the connection among self as well as other people since it is how we, as people, comprehend what others are encountering as though we were feeling it ourselves.

Empathy goes far beyond sympathy, which may be considered a 'feeling for' somebody. Compassion, in contrast, is 'feeling with' that individual, usually by using a creative mind.

Components of Empathy

There are five key elements of empathy.

- Understanding Others
- Developing Others
- Having a Service Orientation
- Utilizing Diversity
- Political Awareness

1. Understanding Others

This is maybe what the vast majority mean by 'empathy':

"sensing others' sentiments and points of view, and taking an active interest in their concerns." The individuals who do this:

- Tunes into emotional prompts. They listen well, and focus on non-verbal communication, getting simple prompts subliminally.
- Show affectability and comprehend others' points of view.
- It can help others dependent on their comprehension of those individuals' needs and emotions.

All these are skills can be developed if you wish. A few people may turn off their emotional antenna when they want to abstain from being overwhelmed by the feelings of others.

For instance, there have been various embarrassments in the UK at the National Health Service, where medical attendants and specialists have been blamed for not thinking about patients. It might be that they were so overexposed to their patients' needs, without having anyone to turn to for help, that they shut themselves off. Mainly to protect themselves from the fact they were unable to adapt. This may have been a paranoid fear.

2. Developing Others

Developing others implies following up on their needs and concerns, and helping them to create to their maximum capacity. Individuals with abilities right now:

- Praise and applaud individuals for their qualities and achievements, and give useful input intended to concentrate on the best way to improve.

- Give tutoring and training so they can help other people, in turn, to develop to their maximum capacity.

- Give extending assignments that assist groups with development.

3. Having a Service Orientation

This means having the ability to focus on the circumstances of the job at hand. It also implies putting their client's needs first and being open to ways that improve their client's quality of life and ambitions.

Individuals who embody this methodology will 'go the additional mile' for their clients. They genuinely comprehend their clients' needs and make a special effort to help meet them.

They become a "trusted advisor" to clients by building up a rich community of connections composed of client and association. This can occur in any industry and under any circumstance.

4. Utilizing Diversity

Utilizing diversity implies having the option to make and create openings for various types of individuals, and acknowledging praising individuality as a way to enhance a community.

Leveraging diversity doesn't imply that you treat everybody in the very same manner; instead, it focuses on tailoring how you connect with others to meet the needs of the group.

Individuals with this expertise regard and relate well to everybody regardless of life experience. When in doubt, they consider diversity to be a chance for understanding that various groups work far superior to progressively homogeneous groups.

Individuals who are good at utilizing diversity have the added benefit of challenging intolerance, inclination and stereotyping when they see it. This creates a climate that is aware of everybody.

5. Political Awareness

Numerous individuals see 'political' abilities as manipulative, however in its best sense, 'political' means having the ability to detect and react to a gathering's enthusiastic propensities and the power in their relationships.

Political mindfulness can assist people with navigating organizational relationships adequately, permitting them to accomplish something where others may have failed.

Empathy, Sympathy, and Compassion

There is a significant difference between sympathy, compassion, and empathy.

Both sympathy and compassion are similar in they both are concerned with the feeling you have for somebody: seeing their misery and understanding that they are suffering. Compassion is built on the understanding of suffering, and they share components; however, the basic understandings of the words are equivalent.

Empathy, conversely, is when you have experienced the same emotions someone else if feeling. It is as if you are experiencing their pain for yourself like you were that individual. This is done through the capacity of our imagination

Three Types of Empathy

Psychologists have distinguished three kinds of empathy: cognitive empathy, emotional empathy, and compassionate empathy.

- Cognitive empathy is understanding somebody's thoughts and feelings in a sound, as opposed to, enthusiastic sense.

- Emotional empathy is otherwise called emotional contagion and is 'getting' another person's emotions, with the goal that you feel them as well.
- Compassionate empathy is understanding somebody's feelings and deciding to take the appropriate steps to help.

CHAPTER 1
EMPATHS, EMOTIONS, AND HEALTH

You should seriously think about yourself as being an empathic individual. Yet, there's a contrast between having empathy and being an empath (a highly sensitive individual who effectively assimilates others' emotions, energy, and stress). Having empathy implies your heart goes out to someone else who's encountering euphoria or pain. Paradoxically, empaths feel others' feelings and physical side effects in their bodies, without the standard protections the vast majority have. Do you think about the hazards of being an empath?

What Does the Research Say?

One of the hazards of being an empath is that you sense more pain. In an examination at the McGill Center for Research on Pain at McGill University in Montreal, researchers found that when individuals with the highest empathy scores were presented to a rise in temperature while they were watching another person experience the equivalently uncomfortable stimuli, the subjects experienced more noteworthy physical vibes of pain than those in the low empathy group. Empaths feel things first; at that point, they think (about them), which is something contrary to how the vast majority work. Empaths sense others' feelings in their bodies without the usual filters; they can hear what they don't state.

It is expected that one out of every five individuals can be considered exceptionally sensitive, and a significant number of these people are also empaths. Some of them can experience social anxiety; however, it is more an outcome than a reason for indications. In empaths, the brain's mirror neuron framework — a specialized group cells that are answerable for empathy — are hyperactive.

Because of this neuronal hyperactivity, empaths ingest others' feelings, energy, and emotions into their bodies. It's alternate wiring of the neurological framework. Being an empath positively has its advantages, including having a more noteworthy instinct, empathy, inventiveness and a more profound association with others. In any case, increased affectability comes with its difficulties, for example, getting effortlessly overpowered, over-invigorated, or depleted, or retaining pressure and negativity from others. Given these risks, it's unsurprising that empaths are especially helpless when faced with gloom, anxiety, emotional burnout and addictions.

Some empaths attempt to numb their sensitivities with alcohol, drugs, nourishment, sex, or shopping. It's extremely normal — being an empath is regularly a missing piece to addictions. If you're an empath, one of the keys to securing your physical and emotional prosperity is to abstain from unnecessarily absorbing others' influence and negative energy. There is a wide range of strategies that can help right now. Your most solid option is to investigate and sees which ones work best for you.

Figure out how to Set Boundaries

One of the risks of being an empath is that others will deplete you. To fight this, limit the amount of time you interact with others in discussions. Recollect that "No" is a complete sentence. So don't be reluctant to state, "Unfortunately, I don't have the time or energy to talk right now," or, "I'm not in the mood for going out this evening because I'm tired." By doing this, you truly secure your energy, so you don't keep on giving to the point of exhaustion.

Question Your Emotions

Because of the risks associated with being an empath, it is vital that you question your emotions. At the point when you feel the unexpected shift of your state of mind or the beginning of emotional exhaustion, ask yourself whether this new inclination is really yours or whether it is better accounted for by another person. "If you didn't feel restless, discouraged, or depleted previously, in all probability, the anxiety you're presently feeling is originating from another person. Perceiving this can help disseminate the feelings you absorb from others — or keep them from having as profound or depleting an effect as they would have had.

Plan Alone Time

Empaths need to invest in finding time alone to be able to focus on themselves and rebuild their energy stores. This time can be spent sitting unobtrusively. You can inhale gradually and deeply. You could focus on your thoughts, or tune in to some soothing music. Remember that submerging yourself in water — for example taking a shower, sitting in a hot tub or swimming can truly quiet you and help expel poisons from your body. With whatever solo activity you pick, the objective is to diminish outer incitement from others and establish and reconnect with your inner sense of self.

Invest Energy in Nature

Being in a fresh, flawless, new condition, like being in the water, removes negativity. It causes you to shed others' energy and renew your own. To determine the utmost advantages, employ all your faculties to experience the sights, sounds, smells and physical impressions (think the countryside, the grass, sand, or soil) that surround you as completely as possible.

Make Real or Imagined Distance

With individuals who consistently drain your energy, don't hesitate to avoid physical contact. Energy is moved through touch, so avoid embracing someone physically and instead send them love from a distance. At the end of the day, giving embraces, holding hands and being intimate is a decision, and it's your right to step away from somebody whose emotional energy is upsetting you.

Thus, you can utilize your visualization powers to isolate yourself from others' toxic emotions. For instance, you can imagine an invisible shield rising around you that blocks another's negative feelings. Then again, you could imagine an invisible band reaching out from your middle to the next person's. At that point, imagine cutting that band to prevent the anxiety or anger from turning into yours. In both instances, you'll still remain present with the other individual. However, you'll be dealing with your own emotional needs.

How to Stop Absorbing Other People's Distress

Is it true that you are a highly sensitive person? Do you relate to the emotions and mental states of others? Does being in a group leave you feeling depleted and fatigued? Is it hard to watch violent movies or even the news? Do you absorb the energy around you and feel what another person is feeling? Assuming this is the case, consider that you may be an empath.

Being an Empath

If you are an empath, you can effortlessly relate to and experience another's sentiments. Empathy can be an excellent quality because it can assist you in interacting with others through an intensive level of comprehension. Most likely, you have great instincts, are a quality nurturer, and radiate healing energy. Being around cheerful, energized individuals invigorates you and causes you to feel great. The annoying part of being empathic is that you also pick up the negative energies and sentiments of bitterness and hopelessness. This can cause issues with anxiety and depression when it is difficult to isolate yourself from the upsetting feelings of others. If you are sensitive to others' emotions and feelings, there are things you can do to shield yourself from absorbing the entirety of the negative energy around you.

1. Name the Feeling

The moment you become sensitive to others' energy, it is difficult to determine whether what you are feeling belongs to you or another person. Naming the feeling when you encounter it can help. If you were having an amazing day and somebody comes along, and you become angry, this impulse probably isn't yours. Having the ability to identify and name your feelings, can allow you to recognize which emotions are yours and those that are of another person.

2. Ground Yourself

If you notice you are retaining the energy of people around you, immediately ground yourself. Concentrate on a particular object that is in your vicinity. Name the things that are around you. Touch something with an interesting surface.

Take some deep, purifying belly breaths. At the point when your attention is on the experience as opposed to foreign energy. It is simpler to keep the environmental energy that surrounds you separate from your own. Practicing mindfulness and meditation can be useful also. The more you can be completely present at the moment, the more prepared you are to assume that the outlandish emotions you are feeling are actually the negative emotions of others.

3. Be Self-Aware

Self-awareness is significant when you are sensitive to others' emotions. Recognize your need for alone time. Invest energy in your feelings. Permit your sentiments to be without judgment. Realize what triggers you in a negative way. Additionally, become mindful of what brings you feelings of satisfaction. When you become increasingly mindful of your sentiments, states of mind, and triggers, it is simpler to tell when you are picking up on another person's energy. Being increasingly mindful will empower you to develop various approaches, help you to recognize feelings that don't belong to you, and avoid absorbing those that don't belong with you.

4. Visualize a Glass Wall

There are various techniques that you can use to keep others' energy independent of your own. One system incorporates envisioning a glass divider between yourself and the other individual. The glass divider permits you to see the other individual's feelings, but the feelings can't enter through the divider. If they hit the glass wall, they are forced to bounce back to the individual and not to you. You can see and recognize the sentiments, yet you don't absorb them. This system also works with large groups as well. You can imagine yourself encapsulated by a glass divider as you travel through the group. Even though you may see their energy, you don't need to take it on as it can't traverse the divider.

5. Be Curious

When it becomes easier to cope with the feelings of another person, you may also notice when something affects them. Regardless of whether your suspicions are right, some of the time, individuals simply need to be heard. Being interested in the other individual and what they are feeling and how it is influencing them can enable you to isolate what they are feeling from what you are feeling. Rather than taking on another person's terrible state of mind, you can ask them questions about what they are experiencing. Increasing a more profound comprehension of what the other individual is going through and why, can keep your own emotions independent, regardless of how empathic you may be. Demonstrating interest in what somebody else is experiencing empowers them to start processing their experiences through sharing them, which can help you both feel closer.

6. Have Strong Boundaries

If you are sympathetic, it is critical to have strict limits. Since it is normal for you to comprehend and sympathize with others, you may inadvertently turn into a dumping ground for their negative feelings. It is important that you know yourself and what you are willing and unwilling to deal with. You may require more alone time than others to feel empowered. Certain individuals and circumstances may be especially depleting for you, and you may need to stop your conversation with them. Strong boundaries assist you with setting limits based on your own needs, emotions, and energy levels. Along these lines, you can keep your presence for those individuals that reliably channel your energy.

7. Release the Emotion

Regardless of whether you consolidate the entirety of the above mentioned, if you are naturally sensitive to the feelings of others, there will be times when you will still absorb them. When this occurs, you can utilize another visualization technique to shield you from becoming overwhelmed. To do this, imagine leaves floating down a stream. Envision yourself recording the inclination you have consumed onto one of the leaves. As your leaf floats down the stream, the inclination goes with it, farther and farther away, and leaves you only with the serenity of the streaming water. Attempt this whenever you absorb another person's feelings and check for yourself whether it is useful.

Empathy is a gift that encourages you to associate with others. The way of being empathetic without negative reactions is to maintain a solid feeling of self. If you end up retaining the negative energy around you, attempt a portion of the thoughts above to check whether they work for you. That way, your empathy, profound degree of comprehension and healing energy will radiate through.

CHAPTER 2
THE DIFFERENCE BETWEEN INTROVERTS, EMPATHS AND HIGHLY SENSITIVE PEOPLE

People regularly lump introverts, empaths and highly sensitive individuals together. Although they share some comparable characteristics, they're each very unique. So, what is the simple distinction — and do you see yourself fitting into at least one of these classifications? Let's investigate.

Introverts

There's been a great deal of thought put into the subject of introverts over the previous decade, and a great many people currently comprehend that being a loner doesn't make you shy or anti-social. Indeed, many introverts are social individuals who love investing energy with a couple of dear companions. In any case, introverts get depleted rapidly in those social circumstances and need a lot of time alone to revive. That is the reason introverts frequently want to remain in or invest energy with only a couple of individuals as opposed to a significant group.

Being an introvert is hereditary, and it includes contrasts in how the brain forms dopamine, the "reward" compound.

Individuals who are conceived as loners don't feel as compensated by encouragement, for example, during gatherings or while talking, and thus they get quickly exhausted in those circumstances. In addition, numerous introverts take profound fulfillment from important exercises like reading, creative hobbies and time for calm reflection.

If you're a highly sensitive person (HSP), you're substantially more likely to be a thoughtful person. Around 70 percent of HSPs are also introverts, so it is easy to see why they're frequently mistaken for each other.

In any case, you can be introvert and not be exceptionally fragile. This would look like being less "in tune" with individuals (for HSPs, the most brilliant thing on their radar is others!), just as being less worried by specific types of stimulation. For example, time pressure, violent movie scenes, repetitive noise and so forth —even though you need a lot of time alone.

Additionally:

- Around 30 to 50 percent of the populace are introverts
- A few introverts are neither empaths nor highly sensitive people
- Introversion is a greatly examined character attribute that is unique to both of the others

Empaths

Empaths are individuals who are incredibly mindful of the feelings of everyone around them. To an empath, this doesn't merely mean a desire to see others' sentiments; the experience is one of being really engrossed in their feelings. Maybe you're feeling their feelings with them. At the point of being overpowered by distressing feelings, empaths may encounter panic attacks, depression, chronic fatigue and physical symptoms that oppose traditional medical diagnosis.

For empaths, this capacity is both a blessing and a curse. It tends to be hard because many empaths feel that they can't "turn it off," or it takes them years to create approaches to turn it down when required. Therefore, an empath can end up going from feeling excessively happy to being overwhelmed with stress, nervousness or other sentiments, and all because another person strolled into the room.

Simultaneously, an empath's capacity to assimilate emotions is their most prominent quality. It permits them to understand and associate with others. This capacity is also what makes them uncommon overseers, companions and partners — particularly when others acknowledge and value their gift.

Like HSPs, empaths also have finely-tuned faculties, substantial natural capacities and require time alone to decompress.

- Empaths can be introverts or extroverts
- "Absorbing" feelings no doubt occurs when inconspicuous social/passionate prompts are picked up and afterwards disguised — an unconscious process that empaths can't control
- The majority of empaths are likely to be highly sensitive people

Highly Sensitive People

Highly sensitive people are regularly misjudged. It's entirely expected to think of "sensitive" as an undesirable thing, which implies that HSPs, in some cases, get unfavorable criticism. However, being a highly sensitive person, you process more data about your general surroundings than others do.

For HSPs, that implies:

- Handling things profoundly and seeing associations that others don't notice
- In some cases, turning out to be overpowered or overstimulated because your cerebrum is managing an abundance of info (particularly in profoundly invigorating situations)
- Taking on emotional signals, like empaths, and feeling a profound level of compassion for other people

- Noticing little and subtle things that others don't see

In effect, highly sensitive people have emotional features, and most HSPs would qualify as empaths — they will, in general, feel the feelings of others like empaths. Also, being an HSP includes being progressively sensitive to all tactile information and not simply feelings. HSPs can become in excessively boisterous circumstances, regardless of whether there are explicit feelings to manage or not.

Like introversion, high affectability has been considered fully. It's to a great extent hereditary and includes a few one of a kind contrasts in the brain. Additionally, it's a common trait shared by up to 20 percent of the population.

- HSPs can be introverts or extroverts
- Most (if not all) HSPs are also empaths
- Empaths and HSPs are different sides of one type of person, but there are more empaths

The Opposite of an Introvert, Empath, or HSP

An introverted person is different to an extroverted one. Extroverts are said to get their energy from society. They have a longer "social battery" than introverts. Their brains are wired to become fulfilled by society.

Narcissism is also thought to be different to sympathy or high affectability. This is false. Being highly sensitive or empathic is good, but so is being less so. Less sensitive individuals aren't affected by changes. Being highly affective can be an advantage or disadvantage, depending on the circumstances. This is especially so in noisy and demanding situations like the office or in the military. These people are not considered narcissistic or selfish.

Many characters with a variety of characteristics exist. This is how it should be. Introversion, sympathy and high sensitivity are significant and invaluable qualities. When we have a society with a wide range of ideas, then the human species thrives. Everything relies upon the circumstances you encounter in life and how well you can utilize your personality and its natural strengths.

Managing Highly Sensitive People

Yvonne works with a small, thriving team in a global marketing company. So she's amazed when a colleague plans a meeting with her to examine his high-stress level.

During their meetings, her colleague explains how rushing through his errands causes fatigue — taking on too much work and changes in general set him off. In some cases, he admits that he feels worried about his condition. Flashing lights, changes in temperature, noise levels and people also make him worried.

He understands that managing colleagues who have unique reactions can be difficult. When Yvonne asks him if he might be a "highly sensitive" individual, he is satisfied. He is also satisfied when she asks him how she can help him get better. They talk about different approaches he could take, like shortening his outings, decreasing stress, and breaking up his errands into smaller chunks. Yvonne likes how competent and responsible her colleague is and enjoys having him in her group. She recognizes some group members will feel more intensely and have more sensations.

Right now, think about what it means to be "highly sensitive." Then we can look at how you can identify people with this characteristic. To do this, we will investigate some basic methodologies you can use to identify these individuals. Knowing this will increase profits and draw out the best insensitive colleagues.

What Is a Highly Sensitive Person?

Highly sensitive people, or HSPs, have a sensitive nervous system. This means they absorb and process more information than the average person does. They also think about information more. HSP traits can be mistaken for the following shortcomings: being unsociable, weak, frightful, neurotic, wretched or anxious. On the contrary, HSPs are frequently found to be extraordinarily proficient, persevering and mindful. They are individuals who are simply more sensitive to their condition and others' emotions than the rest of us.

Step by step instructions to identify a Highly Sensitive Person

It's not always easy to pick out an HSP. Many people are unaware they have this trait. HSPs assimilate more information from their environment than others and break it down.

These means are iterative, so once you've finished them all, you rehash the procedure and refine your perceptions, translations, and mediations until you're happy with the arrangement.

Individuals with sensory processing sensitivity feel more worried about their work than most. However, their managers rate them as the best entertainers. Albeit, an excess of sensory or social incitement can overpower HSPs and make them focused. They are reliable, inventive, persevering and committed people.

This makes them hyper mindful of their environment, and especially sensitive to encouragement. For instance, HSPs can become overpowered when their environment is excessively noisy, extravagant or cold. They become anxious in larger gatherings, multiple conversations, disorder, and mess. They are exceptionally mindful of others' mindsets and sentiments, and regularly understand people's feelings.

These people feel empowered by being around individuals, but at the same time, become overwhelmed by too much excitement.

Advantages of Highly Sensitive People at Work

Chances are, somebody in your group or at your work is overly sensitive. Numerous directors struggle to see HSPs' potential. They can be calm and non-confrontational in nature. However, they can be an extraordinary resource for your group. Here are a couple of reasons why:

- Mindfulness. An HSP's affectability permits that person to see nuances and interruptions in their environment. This makes them mindful of what works and what doesn't, both for themselves and other people.

- Intelligence. These people know about potential "people problems" before they become genuine, and have the understanding to realize how to manage them.

- Empathy. HSPs are regularly intuitive and empathic, and they get individuals, and their thought processes profoundly. This implies they can decipher and resolve relational issues adequately. HSPs hate conflict, and they care about others' sentiments and requirements, which permits them to make friendly workplaces.

- Consciousness. HSPs will, in general, be persevering, cautious and careful about quality. They can see the subtleties and the 10,000-foot view, and they can imagine various prospects.

- Talented. Exceptionally sensitive individuals can frequently be skilled, innovative, keen and fantastic communicators.

The most effective method to Manage Highly Sensitive People

We should investigate six methodologies that you can use to encourage your over-sensitive colleagues, lessen their feelings of anxiety and keep them alert.

1. Acknowledge Highly Sensitive People

It is tempting to try and help an HSP in your group. You have to conquer their affectability regardless of your well-meaning goals; frequently strategies used cause an HSP to feel embarrassed, dismissed, lacking.

Distinctive HSPs are sensitive to a variety of things. They can't change their triggers. For instance, loud noises will cause suffering for some, and too much physical contact affects others. Relax your work environment culture for your over-sensitive colleague. This ensures that you're open, responsive, and understanding of an HSP's situation. Strive to create a supportive and positive work culture. In addition, be mindful that an HSP can appear peaceful. Don't let their manner impact your evaluation of their performance.

2. Address Sources of Stress

Ask your highly sensitive member what overpowers or disturbs them. For instance, ask what irritates them. Maybe it is a murmuring fan. Think about decreasing the attendance of large gatherings and addressing office gossip. Try to address these issues at once instead of making excuses.

HSPs care about their work, but they can also be sensitive to evaluations. Consider giving them positive criticism in addition to any negative feedback. Make sure to emphasize that you value their attributes and that you appreciate their contribution to the workplace.

3. Let People Work Alone

HSPs are also introverts. This means they accomplish their best work alone. Permit your highly sensitive team to work alone at every possible opportunity. Plan for frequent breaks during meetings and gatherings to allow them time to recuperate. Along these lines, permit your highly sensitive team to deal with their own at every possible opportunity, and plan for frequent breaks for them to revive during cooperation or gathering occasions.

As HSPs are profoundly mindful of their condition, they will, in general, feel awkward and will not be able to work efficiently while you're watching them, micromanaging or putting them in the spotlight. They may also see updates or "checking in" as an absence of trust. Give your over-sensitive colleague space to work alone, and clarify that you're available when they need support.

4. Offer a Quiet Place for Work

Offer your highly sensitive team member a quiet workplace whenever possible. This could be a calm place in the workplace or a meeting room, or you could permit them to telecommute. They may request quiet time before anything else to plan for the afternoon.

Urge your HSP to take regular breaks during the day, particularly after a meeting. A day of meetings, occasions, or systems service will negatively affect a highly sensitive individual's wellbeing and prosperity. Allow them to take the opportunity to recharge their energy alone between get-togethers.

You will help their productivity and enable their creative thoughts. Their work can be a profit for your workplace.

Other colleagues may accuse you of giving special treatment. So, make it a priority to treat everybody the same. Make accommodations for individuals that ask for it, when conceivable.

5. Warn in Advance

HSPs' avoid overstimulation by preparing, or creating schedules, plans, and methodologies for upcoming occasions.

Give advance warning about expectations, changes in schedule and other notifications before meetings and exercises. If they become irritated, allow time so they can gather their composure.

Life-Changing Tips for Highly Sensitive People

Highly sensitive individuals often appear weak or broken. Feeling stress or negative isn't a shortcoming; it is a common trait of human beings. It isn't the delicate individual who is broken; it is society's understanding that has become useless and severely weakened. There is zero shame in communicating your genuine emotions. The individuals who are now and again portrayed as being 'excessively passionate' or 'confused' are the very essence of a progressively insightful, mindful and empathetic world. Never be embarrassed to let your emotions, smiles and tears show.

This can be difficult sometimes to remember because it is confusing. Why do you get overwhelmed by assignments that others accept? Why do you think about insults that should be forgotten? Why are the details amplified for you, but lost on others? Remember that you were brought into the world with missing a defensive layer of skin.

You attempt to shroud it, numb it or tune it out. Yet despite everything, some remarks offend you, for example, "You're overthinking things!" or "You're too sensitive," and "Toughen up!"

You're left thinking about everything that is wrong with you.

During my mid-40s, I unearthed the term: "highly sensitive individuals". I was inspired to find out more. It feels great to be one of the thousands saying, "You do that? Me as well!"

Since then, I've discovered that many sensitive people feel disconnected from others. They feel misconstrued and unique and don't always know why. They simply don't understand that they have a straightforward characteristic that explains their behaviors and eccentricities.

There's even a logical term for it: Sensory Processing Sensitivity. These people think profoundly. Some successful people were over-sensitive, such as Albert Einstein, Martin Luther King, and Steve Jobs.

However, when we don't understand how to deal with our affectability, there are consequences. We find it difficult to even think about keeping up with other people. We attempt to do what others appear to deal with effortlessly and attempt to show that we are just as good. This can create conflict.

We work hard and do a good job and use our natural gifts. We are imaginative students, upright representatives, and committed relatives. Sometimes we feel guilty about the past, but it can be damaging. The damage shows up as changes in our wellbeing, muscle strain we can't dispose of and feeling perpetually exhausted or nervous for no reason.

If any of this reverberates with you here are ten actions you can make to stop battling and begin flourishing:

1. Stop Looking for a Person or Thing to Fix You

Affectability is a disposition quality, not a medical disorder. So, nothing is innately wrong with you. Tragically, wellbeing experts don't fully comprehend sensory processing sensitivity because it is an ongoing region of research.

Of course, highly sensitive people are bound to have hypersensitivities or sensitivities to nourishment, synthetic compounds, prescriptions, etc. Also, they're increasingly inclined to overstimulation and in addition to feel pressure more rapidly. These can prompt other medical problems. In any case, affectability in itself isn't something that requires fixing.

Successful sensitive sorts understand that they're not "broken." If your brain is depleted from hectically inquiring about your endless flaws, you have to learn to accept that. Realize this is the way nature intended you to be.

2. Ensure Yourself Regularly That You Are Not a Cheat

Impostor syndrome isn't just a problem for highly sensitive individuals. Numerous honest and highly accomplished individuals succumb to this annoying fear. But, the building distress about being discovered is always at the forefront of a delicate individual's mind.

Is there a reason you have felt unique your whole life and unable to fit in? Perhaps during a soppy movie scene, you blame your tears on dust in your eyes. You try to say you applied for a job only for the fun of it. It can be embarrassing to feel ashamed of your body and mind. Don't let society force you to camouflage who you are.

Successful sensitive types claim that their sensory systems are wired uniquely in contrast to 80-85% of individuals. Maybe you continually ponder who you ought to be or what you ought to do. Instead of focusing on this, be proud of your accomplishments. Find the qualities that show yourself as you genuinely, even if you feel like an odd ball.

3. Search out kindred spirits (and realize that you are not the only one)

You presumably feel different and alone. Yet, in actuality, you're not. Many have encountered disarray in confinement before finding out there are others like you. They've felt the flood of intensity that originates from being upheld by similar spirits. Also, show proactive kindness.

The key is to spend time with other sensitive individuals. Look for ones that have successfully accepted who they are. They are the ones that comprehend how to deal with their affectability. They know how to employ their superpowers. They understand what it means for you to feel perpetually under attack, and they can offer first-hand understanding and insight to assist you with making your sensitivities work for you.

Effective sensitive sorts acknowledge and relish the qualities of affectability. They do this for themselves and other people. If you're feeling unsupported or misconstrued, discover a delicately learned mentor, guide or network who understands you and support that association.

4. Search for The Hidden Positives in Each Circumstance and Absorb It

The cerebrum is an incredible channel that molds encounters and the impression of the real world. If you think of the world is a hazardous spot, your cerebrum is wired to search for proof of a threat. If you trust think of the world as a kind, caring place, you will notice the more positive things about your world. Whatever you concentrate on will be what you tend to get more of.

As a highly sensitive individual, the more negative your nature, the more you will endure. Be that as it may, the inverse is likewise valid — the more positive you are, the more you will flourish (even contrasted with others).

Thoughts are improvements for your sensory system. One

of the most important things a sensitive individual can do is recognize the negative. Don't disregard it because that will only create conflict. Let go of oppositions and inundate yourself in positive considerations and circumstances that cause you to feel great. Make yourself liberated.

Successful sensitive sorts choose to see the world overflowing with chances to feel thankful for and to absorb that positive vibe. When you feel helpless, remember that your contemplations (and the enthusiastic charges they trigger) are always in your control.

5. Find New Spins on Old Imperfections

Your gifts of affectability incorporate profound reflection and intuition to see all points and results. However, you are also constantly thinking about your faults. You are easily overwhelmed and depleted by constant excitement. It is difficult not to see your shortcomings as blemishes. You may not always understand why you feel and carry on in the way you do.

In truth, these "weaknesses" are your neglected needs and enjoyable blessings to support. In reframing your past and sustaining your present, you set yourself up for accomplishment in your future.

Successful sensitive sorts evaluate old discernments. They learn to refine their understandings of affectability. Search for the opposite side of the coin, especially when you feel overwhelmed, easily affected, dismissed or detested. You'll discover a portion of your most unique qualities: instinct, vision, honesty.

6. Treat Yourself With Sympathy

As a highly sensitive person, you are profoundly empathetic, to such an extent that putting others' solace and needs before your own is natural. What's more, you're frequently your own greatest enemy. You propel yourself hard, and afterwards, you beat yourself up when you come up short. You scrutinize yourself in ways you'd never imagine for others.

Controlling your annoying inner critic is basic to self-sympathy. Don't disregard your inner critic, but don't dwell on it. Grasp your powers and consider your profound thinking as a blessing. Take control by hearing your musings without judgment. There are pearls inside your thoughts. Afterwards, contemplate how to spread kindness. Cherish the feelings you have. Pick activities that promote positive thinking for yourself and others.

Successful sensitive types show themselves and others equal adoration and sympathy. It might feel narrow-minded or vain from the outset. However, it's most certainly not. If your basic internal voice is cheapening what your identity is, answer back with self-generosity. This is the antidote.

7. Make Sound Limits, not Rigid Emotional Walls

Our culture tells us to "take a painkiller and push on" unquestionably more than its esteems affectability. We grow up hearing: "no pain, no gain; natural selection; life isn't reasonable — get used to it." We appreciate individuals who demonstrate determination to overcome their difficult circumstances.

As a highly sensitive people, your reflex response might be to freeze up or battle to toughen up. You fabricate dividers to shield yourself from hurt. These can include emotional dividers. Smothering sentiments or making strife sensational diverts you from the real reason for your torment. One physical divider is heaping on layers of weight to hide behind. Mental dividers, for example, can be blocking out pain with alcohol or medications.

Or then again, you may let every one of your limits breakdown without a moment's delay. You unknowingly retain others' energies and feel burdened by capricious occasions and feelings. You attempt to get away from the sentiments by becoming involved with over-thinking everything. This includes perpetually arranging, looking and breaking down. This creates a divide between you and your instincts. Furthermore, in the process, you mistake good faith for exhaustion, sympathy with over-tolerance, and empathy with over-resistance. Stop beating yourself up about thinking you ought to have better limits. It's an endless loop.

Successful sensitive types set limits. If you struggle to put your own needs first (which is not usual for an exceptionally sensitive individual), settle on just saying "no". Practice doing this. Saying no can be done with kindness and effortlessly. Carve out alone time in your day to energize. Choose to do good things for yourself.

8. Take Notice of the Signs Your Body Is Giving You (resist wavering between emotional extremes)

Many susceptible individuals try to overlook the messages their bodies are sending them. They switch off to keep from becoming overwhelmed. They focus on the needs of others rather than their own. Does this sound like you?

Doing so leaves you swinging like a pendulum. You go from one extreme to the other. Sometimes you swing quick and other times slow. This rocks you from between being over-animated and mind-numbingly exhausted eating less junk food and afterward gorging, or to practicing hard and afterwards requiring a few days to recoup. There are so many examples.

Successful sensitive types take notice of the physical sensations in their bodies. They know their body is sometimes disagreeable but they trust their bodies to direct them. If you have a propensity for avoiding emotions or becoming overpowered by emotions, figure out how to understand your body's messages. You'll invest less energy in being persecuted.

9. Design Healthy Habits That Fit Your One of a Kind Needs

In the end, everything finds you. Overwhelming hours grinding away, trailed by hard sweat at the center, and keeping up your home. These are all energized by a negative perception. Your weight concerns control you. Insignificant rest and less alone time snares you. The vast majority of people are able to deal with these issues. For you, it seems like a trap.

Solid propensities hit a sensitive nervous system hard. This is like being well nourished because you take care to avoid sugar and food additives. Even exercising without giving yourself time to recover can be negative to your health.

If you allow a lot of excitement in your life and don't spend enough time recovering, you risk sickness. In addition, if you overprotect yourself, your efforts will go unacknowledged and result in sickness.

Successful sensitive types perform practices that help them. If you struggle with vitality or prosperity issues, organize propensities that support these parts of your life. You can choose to rest or take some alone time. When you feel like something is overwhelming, take a break. Try lifting weights or engaging in other exercises.

10. Stop Smothering Your Sensitivity

After a lifetime of being besieged by stimuli you will want to push affectability out of your conscious awareness. Tuning out of tireless sensations is one strategy, or imagining you don't care at all. Mitigating serious sentiments, both good and bad, prevents you from becoming stressed. Practice being creative and let go of stifling feelings.

This self-protective mechanism may trick your cognizant mind, yet it doesn't trick your sensitive body. It creeps into your wellbeing, your connections, your profession, and each part of your life. Beware of straining yourself otherwise you will harm yourself.

Successful sensitive types let go of the grip for control. When you let go of negative energy, you free the blessings of affectability. You regain sympathy, innovativeness and increased delight. You permit your actual potential to bloom.

As you're working through the tips above, remember that the way to flourishing is to understand it's okay to be sensitive. Everyone has both difficulties and qualities.

Utilize your creative mind to perceive shrouded

understandings, and purposely pull together on energy and potential outcomes.

Utilize your deep-feeling body to tune into your feelings and sensations. Remain inside your ideal scope of excitement as much as you can.

Utilize your heightened awareness to move to whatever beat you want.

You will learn that others will hit the dance floor with you.

CHAPTER 3
EMPATHS, LOVE AND SEX

Is it true that you are fascinated with an empath who confuses the hell out of you? At one moment they are upbeat and energetic, the next they are crying and need to distance themselves. My ex Matt would say, "Colette, you're excessively hot and cold. I don't have the idea of how to manage the to and fro."

Matt isn't an empath. And keeping in mind that he adored me, he was ignorant of my needs. I expected to feel cherished.

Empaths make up around 15–20% of the populace. We're an uncommon breed. What makes us different is our hyper-responsive mirror neurons. Empaths are not the only ones who deal with feelings of helplessness. Others react too. But empaths have additional reactions. We react to the energy of situations and substances the same way everyone reacts to nourishment. Nobody has shown us how to process all our emotions successfully. People say it is terrible to be oversensitive. They try to suggest we have misjudged the situation. Most of the time an empath accepts there must be something terribly wrong with them.

"Hot and cold? To and fro? You're damn right I am," concludes the empath.

It's called being an empath who feels everything — your feelings, emotions, the energy of the room, the climate and the feelings of animals. Empaths can even feel the enduring of individuals on the opposite side of the world. There are so many examples I could go on forever.

Don't worry if you haven't mastered your sensitivities, you will endure. Keep battling to be comprehended. You will close down and strike back when you're activated. You will have fights and failed relationships.

Your goal is to be recognized and be compensated for the gifts you have. Strive to never again feel like you have a curse.

Instructions for Loving an Empath

Be interested and open. Rather than believing we're nonsensical when we cry at the drop of a hat, ask us what we're feeling. Being curious about what's happening within us will help us to communicate our emotions better.

We've been shown how to smother our feelings our entire lives, which has the negative effect of increasing them.

We become a pressure cooker loaded with fear, disgrace, blame and anger. That crap will detonate all over you, getting in your hair, on the roof, up to your nose, and you'll be left to tidy up the wreckage for a considerable length of time. This is just one example of what happens if our feelings are ignored.

At the point when you are open and welcoming, we have a

sense of security to tell you about ourselves. This causes us to feel seen and heard. We truly need you to get us. Your interest will mollify us and ease our uneasiness.

As opposed to thinking we are hot/cold or indecisive, you will see we are perplexed and profound. You will cherish us considerably more for our uniqueness.

If you have a closed brain and body, you will be disappointed. Consequently, we will feel your dissatisfaction.

We will close up unless acknowledged. Furthermore, what used to be fun and exciting (hot) will become grim and quiet (cold).

Heart-based Communication

Figure out how to address us from your heart and not merely from your head. We react to clear and succinct communication; however, that doesn't mean tell me everything you are feeling. We talk about emotions.

Even though you probably won't be as versed in the language of feelings, you are competent.

People wear masks to avoid sharing their souls. Here's a chance to connect with your sentiments and express them to us. Reveal to us how you're feeling. Mention to us what you want and what you fear.

Associate with what's in your heart and offer that to us. You may need to delay and approach things uniquely. That is alright. We're tolerant.

However, we have no patience for manipulation. Stop attempting to persuade us to be consistent. We follow our emotions. We suggest you follow your own too. Real connection, profound closeness and soulful pleasure are three positive qualities. Follow your sentiments and communicate to others what you feel. It is what you are good at. All three of the mentioned qualities are entwined with each other. Just like how you need to be with your friend or partner.

Real connection is fundamental for us to open up and uncover our positive qualities.

We have to feel heard and ultimately observed by you. We need you to realize our situation is unique and not like others. We don't care about the same things ordinary people care about. We need to explore your feelings and find out what lies underneath the surface.

Intimacy resembles air for us. We need it.

Sometimes the topic of sex is difficult for us to address. We long for the closeness created with another individual. For those of us who are comfortable, we appreciate sex.

Not just any kind of sex will do. Sometimes a sexless marriage will force people to leave the relationship. Please consider everyone's feelings in a relationship. Forming trusting relationships is important to us.

If you stick with us, be assured; we will be loyal. Being a part of a couple is something we appreciate. Consider the following story to clarify.

Someone I know was presenting at a workshop attended by 30 people. Most people accepted that joy is experienced by the body. But this wasn't always true, I claimed. Joy can be shallow if experienced only through the body. It must be also experienced with the soul. This includes sexual energy.

Sexual energy isn't equivalent to the sex organs. Sexual energy is a life power, and has potential power. It is the very center of our being.

Many people ignore this part of themselves. They identify with their bodies or their job. But this is shallow and with the view you will never be fulfilled.

So now you realize where to focus. There are things you should stay away if you love a feeler.

Protect us from your feelings.

We don't require reassurance. Let me state that once more. We don't require reassurance, or even saving. We don't want your pity for being unique and sensitive. Avoid telling us about how bad we are. It could trigger us.

If we do get triggered, we have to manage it.

If you want to address an issue, first ask if it's the right time to talk. Offer us the chance to react as opposed to expecting we can't deal with it. Possibly we're feeling pushed to the limit. Let us refuse to talk until we feel more energized.

Don't try to shield an empath from their emotions. Focus on successful communication. We invite this.

Rationalize with us. Stop assuming we are frail and unequipped to deal with our own feelings. Give us time to understand your viewpoint.

Our sensitivities don't make us powerless.

Being an empath is a blessing regardless of whether 80% of the populace doesn't get us

You may imagine that having a hyper-responsiveness to energy and other's feelings is a mishap. Unfortunately, a lot of empaths also think this. Don't let this false impression make you think it's justifiable. It's simply false.

Our empathic capacities resemble superpowers. Like most heroes, we need experience in order to learn to utilize them. That is the reason I love a proper emergency.

We call forward an emergency when we're prepared to learn, recuperate, and develop. The breakdown goes before the leap forward. An emergency is gold.

When we are in an emergency, don't attempt to rationalize us or attempt to fix it for us.

Instead, be interested in what's happening within us. Convey your feelings and try to communicate with us in a profound way. Trust that we can deal with it. With your adoration, we will transform the agony into something helpful. We will heal ourselves.

Call us "too sensitive, too emotional, or too much"

Sometimes a normal workload can be overwhelming for anyone. Somebody who has a normal mirror neuron framework can't comprehend what it means to have a hyper-responsive one. So stop attempting to think for us.

This is what is actually going on. One person thinks, "That person doesn't understand what is going on. They are over-thinking the situation." Meanwhile, your empathic friend is focused on the inconspicuous energy and subtleties you don't perceive. Try to understand so you can understand the empath.

In this situation it might be you who is "behaving in an extreme way".

Empaths have heard these criticisms their entire lives. They have figured out how to stifle their feelings instead of clearly communicating them. The goal for the non-empath is to avoid enabling that behavior. Help them to set their mind and body free.

You can help mend the injuries that empaths need to manage.

You need to accomplish your work to be a successful

human. There is an advantage over figuring out how to adore yourself and your companions.

If you find yourself interested in an empath and want to start a relationship, let me offer some pointers. Talk to them in private. Investigate them from the inside out. Help them help you. Doing so will transform both of you. It is the impetus for self-healing.

Empath's have a special way of communicating. You can learn it. When you do this will open up an entirely different world for the both of you. A mysterious world full of emotions, closeness, and profound joy. These things are what empaths need from others. Use these techniques to mend any issues. Help them to get over past injuries that they still suffer from. If you can do this, the empath will reward you. They will give you their heart and always be there for you.

How to Know If You Are a Sexual Empath

Regardless if you are single, dating, or in a long-term relationship, sex matters. It doesn't matter if you are an empath or a sensitive person.

In this book, we have been discussing how to survive with an empath. Empaths are sensitive and so there is nothing casual about casual sex. During lovemaking, empaths can get both nervousness and pleasure from their partner. They pick up on their partner's instincts and sentiments. Advice for the empath: pick your partners carefully. Sometimes lovemaking can become negative and lead to stress or dread. This is especially valid if you are a sexual empath.

What is a sexual empath? Somebody whose empathic capacities heighten during a sensual experience with the goal that the individual in question detects more pressure or ecstasy. Sexual empaths are highly sensitive during lovemaking. They can also be a tease. They can get a partner's energy considerably more than different empaths can. For all empaths, but particularly the sexual kind, they need a responsive partner to feel their best. The ideal partner would be one who can respond with love and respect.

Shockingly, a few empath patients have committed errors when they've been without a partner for quite a while. When an empath is anxious to start a relationship, they overlook red flags. They end up making poor sexual decisions in the early stages of a relationship. Many times, it is because they haven't been in a relationship in a while and end up picking the wrong partners.

This opens them up to risks such as getting hurt. They find people who can't cherish them or who are inaccessible.

Tantra workshops can help. So can arranging private sessions with a tantric teacher. Tantra is an ancient practice that joins sexuality and otherworldliness using body-focused activities. In private or group circumstances, you will be educated to tune into your body, tap into your sexuality and otherworldliness, and work through old injuries, ruinous relationship examples or deadness that prevents you from feeling. These meetings increment your sexuality and keep it flowing. They boost your forces of fascination as opposed to permitting this energy to die. Others may not feel how provocative you are if that occurs.

When you've discovered a partner, who is completely coordinated with you, consolidate your heart with your sexuality. Empaths flourish. It is magnificent when sex, soul, and heart are consolidated in lovemaking.

One goal of heart focused sexuality is figuring out your partner's boundaries. For example, if your partner had a disappointing day and is irritated, empaths should avoid them. They might absorb their partner's indignation. Have a plan about how to deal with these situations. Otherwise you run the risk of making others angry or feeling stressed.

Teach your partner about your sensitivities. If you are involved with an empath, clarify your questions and answers. They will address your issues. The empath's universe is unique. Persistence will have a significant effect on your relationship.

When Two Empaths Fall In Love

Everybody needs to be appreciated by those they're near. An amazing relationship can happen when two empaths form a bond.

When collaborated with another empath, they'll experience a situation where their other half comprehends them on a basic level. They feel what they're feeling too.

The following names advantages and disadvantages that happen when two empaths pair up.

Pros:

Soul-Deep Understanding

When your partner understands your feelings because they feel what you feel. They comprehend what you are saying on a higher level.

You may have experienced life feeling as if no one else could truly identify with how you experience the world. You are excessively hypersensitive to stimuli and so forth.

There's someone else who can feel what it is your experience. They may be different from you and don't see the world that same way. They see through your eyes, feel through your skin. That is huge.

That is one of the most unfathomable sentiments on the planet.

Common Happiness Is Absolute Bliss

You know when you become energized and glad about something and need to impart that satisfaction to the individual you love? Indeed, when you're both empaths, that is actually what occurs.

What does it resemble?:

Do you recollect that exploding volcano experiment in science class when you were a child? When you combined baking soda and vinegar, and there was a giant explosion? It resembles that, just with energy and enchantments and glittery rainbow butterfly unicorns and stuff.

You'll Have Wonderful Animal Companions Together

One thing that practically all empaths share for all intents and purpose is a profound love of animal friendships.

It tends to be troublesome when an empath is involved with a non-empath because an individual may not see the extent to which it is so critical to have non-human friendship in the home.

At the point when two empaths coexist, it's basically ensured that there will be furred, fluffy or potentially fishy friends there too.

You share your space on the couch with your animal friends. If you are lucky you may even wind up running an animal sanctuary or rescue together.

Wouldn't that be great?

Amazing Care and Nurturing

Empaths put the needs of others before their own. When they are in a relationship together this is magnified.

Consider this: if your partner's joy and prosperity are the highest need to you, and your joy and well-being are the highest need to them, this creates an amazing situation. You know each other's feelings and can determine what each other's needs are.

If you have a cold, your partner will appear with a bowl of soup. They just know you require it.

Surprise your partner with an outdoor excursion or a hike. Time in the forest might be what they require. The empath partner will detect these needs.

You'll envision each other's needs. Sometimes before you even acknowledge them for yourself. A lot of affection and care will flow between you.

The Sex Is Spectacular

Have you ever felt so connected with your partner that you can't tell where you begin and they end? This is a typical event in "empath sex."

Recollect that bit about having the option to foresee each other's needs, and putting each other first. When both parties are centered on the other individual's pleasure, they can detect what they need. They know how to be completely present and hyper-focused on each other. There's nothing like it in the world.

The intimacy that is created when two individuals can

merge energy and feel what each other is miraculous.

Cons:

Bad Moods Are Contagious

Being an empath is magnificent when high-vibrational, "happy" energy is floating around; however, when your other half is battling with troublesome feelings, they can influence you just as strongly.

Not because your accomplice is lashing out at you, but since you can feel what they feel.

Sometimes it goes undetected. Superficially they appear alright. But then one empath will begin to feel furious or on edge and have no clue where these sentiments are coming from.

Feeling a storm of emotions without having the option to perceive where they're beginning can be truly confusing. To recover a sense of tranquility, physical separation is best.

So Is Physical Pain

It's not irregular for empaths to feel empathy pain when those near them are suffering. This is the reason you'll know when it's been a bad day at work or if one partner is thinking about having a baby.

Having the option to feel others' physical risks can be extremely odd. It can be hard to manage, particularly if they have chronic pain conditions like joint inflammation or fibromyalgia. How might you deal with the agony from a condition that you don't have?

Mostly, empaths can look at an individual and have an option to read their history in their skin, their eyes and their energy.

This can be useful if somebody calls a specialist. They might need help to mend. Sometimes relationships can be hard, even for empaths.

We've all experienced troublesome times. That doesn't mean we like to reveal everything about ourselves to others. It can be most trying at the beginning of a relationship. But setting aside some time to develop healthy communication works. It's like an onion that you strip layer by layer. We don't need to reveal everything before we are ready.

If you've experienced some awful crap that you aren't prepared to discuss, and your partner makes reference to those careful circumstances since they "know" you've managed them, your reaction may change from awkward to disturbed.

This is true in the opposite situation. Your partner doesn't feel like they need to tell you about how they manage certain things. For example, their fixations, fights, or a number of sensitive subjects. They want to choose when to open up to you. As an empath you know what they are feeling, but that doesn't mean you shouldn't respect their right to choose when to talk about it.

That can be extremely hard to contend with.

You'll Both Need Alone Time to Recharge

Empaths realize that they need isolation to revive their batteries. It is simpler to resolve issues when both parties are ready. When they are mindful of their own self-care and needs, then an empath is ready to reach out to another. If an individual doesn't know about their empathic nature or is a sort to be clingy and mutually dependent, they may take their partner's requirement for isolation as a kind of dismissal.

Regardless, sometimes people just need space. This can be seen as contemptuous. An empath's partner might think they are not adored or worse seeing someone else.

This is something that should be imparted unmistakably and adequately, with a lot of consolation.

You Can't Hide Your Feelings

Your partner will recognize when you are trying to manage anxiety, sadness or some other troublesome feelings concerning your relationship.

A great deal of us wants to keep our thoughts and feelings to ourselves until we can process them. We need time to make sense of what is happening and decide on our best course of action. Family obligations and work sometimes make this time to reflect difficult.

One advantage of being an empath with an empath partner is having the power to understand each other's feelings. You don't have to work through difficult times all alone.

It's baffling when they don't concede you the space to get yourself straightened out. They realize something isn't right and request that you talk it over with them. Usually the cloud of emotion is influencing them too.

There are various degrees of empathy, obviously, and no two relationships are ever the same. Some empaths will click with each other; some may point each other out as excessively exceptional. Some don't associate enough. That is entirely okay.

CHAPTER 4
PROTECTING YOURSELF FROM NARCISSISTS AND OTHER ENERGY VAMPIRES

Are you involved with an energy vampire? Surprisingly, 20 percent of men and women have vampire qualities or are full-blown Cluster B. That is one out of five individuals. What's more, every single one of them influences five additional individuals. About 60 million individuals knowingly or inadvertently are influenced by energy vampires. That means you or somebody you know is involved with one. This is especially the case if you are an empath or exceptionally delicate individual.

The energy vampire in your life could be a parent, a partner, or even somebody you consider to be a companion. You could be compromised by an energy vampire and not even know it. They are exceptionally beguiling and bombard you with love. That is until they come after you.

Before long, you are sucker punched with insults and being shamed. They chide you for your societal position, body size, age, pay level, how you talk, or where you originate from. They can get physical and harm you. Energy vampires get surly and distant. You begin to tread lightly in their presence and use up your energy to satisfy them. You use a lot of your energy to applaud and show your appreciation just to maintain harmony. This can adversely affect your self-esteem to the point that you think something is innately wrong with you.

Living with an energy vampire causes endless aggravation at a cellular level because of elevated levels of cortisol going through your body. Consequently, you start to make poor dietary decisions or even turn to alcohol or medications. That only furthers cell irritation and causes illness. Many empaths don't understand that an energy vampire is draining the life out of them until they become genuinely sick.

Fortunately, health care providers and society can help by explaining how these energy suckers work. There are recognizable personality traits and common manipulation tactics that identify the energy vampire. At the point when you see these characteristics, you can figure out how to protect yourself.

Ten Strategies for Protecting Yourself against Energy Vampires

If you are involved with an energy vampire, you have to realize how to protect yourself. Self-protection ought to be your essential objective when managing energy vampires. You have to do this even if the energy vampire is a relative, partner, or dear companion. Distancing yourself from these people needs to be a top priority.

Here are a few procedures to assist you with shielding yourself from energy vampires:

Acknowledge they exist

A great many people who are involved with energy vampires are empaths. We empaths accept that everybody is acceptable. But we frequently remain in a dangerous relationship. This strengthens the energy vampires' control. We would prefer not to concede they are genuinely in it for themselves and need empathy so they couldn't care less about you. Ouch! Understand that there are a few bad people. This will enable you to protect yourself.

Keep a gut instinct journal

Empaths are exceptionally intuitive. Be that as it may, after numerous years with an energy vampire, you can lose the capacity to accept what you feel. One way you can begin to trust your gut is to keep a Gut Instincts Journal. Focus on what your gut says about an individual. For instance, does the individual take part in dangerous conduct or compromise self-injury? Does the person in question tell lies, cheat, or have repeating issues with the law? At that point monitor how circumstances play out. Whether your vampire is convincing and enchanting, don't ignore your gut senses about an individual. Make sure to focus on how they treat all of the individuals they interact with. For example, a server at your local restaurant, or a section level worker.

Find a reality check companion

Make certain to have a sensible and dependable contact with "vampire radar" whom you can contact when you are feeling dubious. Frequently this is an old buddy who knows you well and who hasn't been taken in by your vampire. Call this individual at whatever point you're feeling uncertain about a circumstance.

Put yourself first

Energy vampires will battle for control. They can also become angry and manipulative, or frequently forceful. What's more, they are very skilled at "parting practices". That means they are good at setting one individual in opposition with another. At the point when you experience these characteristics, step back or leave the room. Remember that you deserve a blissful life in which your needs and feelings are reciprocated. Promise faithfulness to yourself. You can say, "I vow loyalty to myself and to my spirit for which I stand. I respect my blessings and my ability. I focus on staying faithful to myself from this minute forward for the entirety of my days."

Pat yourself on the back routinely

Most empaths give others an excessive amount of credit and acclaim for their accomplishments. We will talk more about this later. Instead, pat yourself on the back and remind yourself of your identity and admirability.

Say "no."

One of the ideal approaches to protect yourself is to limit your connections with a vampire. You can say you are previously engaged elsewhere. Learn to turn individuals down. Saying "no" takes practice. Also, it's everything about sympathy, which as an empath, you have a lot of. If you see it as too hard to even think about saying "no" from the outset, start by saying, "I'll hit you up." Stop the habit of the automatic "yes."

Get support

Finding out about vitality vampires requires support. Don't think about asking for it from your rude companion. One person to turn to is a psychotherapist specializing in narcissism. They can be helpful at this point. There are also narcissistic group meetings that you can join. Consider a couple's therapist if you are still with your partner. Choose a therapist who knows how to manage difficult personalities.

Organize quality time with yourself

Take 15-20 minutes out of every day just for yourself. Accomplish something that makes you happy. For example, ponder over things, clean the house, or walk in the park. Picture your negative energy being wiped clean from your body. Take the time to process your own feelings and reconnect with your needs. This is the genuine driver of your activities.

Set limits

Regardless of whether it's online life, the news, or social networking, set limits. This is the basics of feeling well and prospering. Empaths need to set limits. Break away from your screen time, read books instead that satisfy you. Try to stay away from the news. You might want to help everyone, it is one of your main motives, but you have to set limits. You can't do everything. If a friend needs someone to talk to, make the time to talk to them if you can. But still set a limit to that discussion. Afterwards return to your own life, including some extra self-care to assist you with relinquishing your energy.

Take inventory of your relationships

The individuals you invest energy in should help you, not sap your energy. If you know somebody who sucks away your energy just through their presence, be careful. You could be managing an energy vampire. Energy vampires don't care about others. This means you won't get anything from a relationship with them. They only care about themselves. Stay away.

As an empath, you are here to think about your being. What's more, the view of an empath is grounded continuously in adoration, sympathy and administration, not altruism and suffering. At the point when you love yourself first, you remain as the light. This will show energy vampires that you are not readily available to be their prey. What's more, you make it more secure for others to do likewise.

How to Block Every Type of Energy Vampire

We've all been there. You just got off the telephone with a self-absorbed relative, or said farewell to a friend at breakfast, or got wrapped up in a meeting with your chatty manager, and you feel drained. You're bad-tempered and prepared for rest, preferably on a desert island where nobody will request that you gesture and tut your way through a new 20-minute tale about their feline's muddled dental, medical procedure. This is the thing that happens when you've been sapped by a vitality vampire: They possess large amounts of the working environment, in your family and your group of friends—and luckily, there are approaches to shield yourself from vitality vampires' life-pith leaching habits.

There are several distinct types of energy vampires. What they share in common is the way their conduct can affect someone else's emotional state. "You can out of nowhere be feeling awful. You may feel depleted, or even genuinely sick," one victim of an energy vampire said. "You just want to eat sugar and carbs." You feel on edge, or angry, or negative when you didn't feel that way before. Or, on the other hand, you feel disgraced, and simply retreat.

Common characteristics of each sort of energy vampire follow. They are in our midst. We need to know how to obstruct their "assaults".

The Narcissist

According to the Diagnostic and Statistical Manual of Mental Disorders, narcissism is "an inescapable example of pomposity, the requirement for value and absence of empathy."

Narcissistic energy vampires are in it to control someone else. They use techniques like giving you the silent treatment or gaslighting you. Those are powerful tools for the vampire.

The most effective method of blocking them

Keep a clear distance between you and the emotional vampire. Many of them are childish and not all of them are narcissists. Some think that they can be healed, but we will discuss that later on in the book. Instead they consider the problems in their life to be the other person's concern. They never own up to mistakes.

However, there's a promise for some who want to change. When an energy vampire wants to do something, the people who love them usually follow. If they want to move out of town, their partner will do it too.

The Victim, or the "Poor Me"

They don't assume liability for themselves. These are the poor me kinds of people. They feel as if the world is against them. They will spend hours on the phone telling you about why their manager doesn't get them, why their sweetheart parted ways with them for the tenth time. They never admit to their faults. When you offer any condolences or suggestions, they tell you they don't need it. Then they call you the next day with the same complaints.

The most effective method to block them:

Energy vampires need someone to talk to. But that doesn't mean you should be their full-time supporter. Don't be their advisor because they will abuse that power. It doesn't help them. Instead recommend them to an advisor that could help them.

The Drama Queen or King

What they do: Ever notice the individual who says they're "finished with all the drama" is by all accounts continually involved in the show? They think they are on a television show. If you know someone like this, you're dealing with a drama queen or king.

Everything becomes a big deal. If they get a spot on their skin, they think they have cancer. They get into a minor mishap, and they say they nearly died. There complaints could deplete a crowd of psychological energy.

An empath dealing with an energy vampire causes problems.

The most effective method of blocking them:

Avoid asking follow-up inquiries.

Don't look them in the eyes. Avoid non-verbal communication. Point your body away from them. Empaths can set clear limits and tell the other person they are busy. It's okay to tell them you care about their feelings.

The Control Freak and the Critic

What they do: The individuals who take on these practices may have good intentions. However, time at work leaves them feeling drained, sensitive and cautious. The critic may tell you that you've put on a couple of pounds. They point out your faults. Controllers are people who need to control everything that you do. They want you to do things their way.

Instructions to block them: once more, limits are your companion. If the energy vampire is a friend or relative, you can tell them they offend you. Controllers need gratitude. Tell them you will consider their needs, but keep your distance and walk away. Try not to get into it, and don't attempt to control a controller.

The Passive-Aggressive

What they do: They communicate in an angry way but have a smile on their face. This sometimes feels like they are sabotaging you. For example, offering a container of sweets when you state you're on a diet or holding you up when you've made other arrangements.

Step by step instructions to square them: Identify each issue they have and try to address it. Show them how their behavior and words don't match. Tell them that if they want your attention, you are more than happy to give it, but they need to stick to an agreed upon time. Many people don't make noise when they get angry and instead, they harbor it. This is unacceptable.

When managing an energy vampire, come prepared

Be prepared to disable them so the same thing doesn't continue to occur. This is defensive self-care. Not all energy vampires should be restricted from your life. Just make sure you have strategies for dealing with them. Get grounded, become clearer, and preplan your words. Rehearse what you will say with a trusted friend.

Emotions can be contagious.

An individual who constantly shares wild suppositions can encourage people around them to take on their pessimistic viewpoint. The good news is certain feelings increase.

Suppose you have an associate who is exceptionally mean or is frightened of being fired. That person tells everyone. They say they are worried they won't have a job anymore. That fear will spread all through the work environment. This is known as an emotional infection. That's when adverse emotions spread. But it can also be positive ones that spread too. Imagine someone saying that they are so happy to be there with everyone. That lifts peoples' spirits.

Emotional vampires aren't an actual monster.

Narcissists are special cases of emotional vampirism. They aren't that much different from each other. Some will want to change. They will want to develop so they can heal. If they come to treatment, they can change. They don't mean to be the way they are. Depleting others isn't necessarily their goal. They lack the ability to be mindful.

What's harder to think about is that none of us are invulnerable. Anyone can be an energy vampire. From time to time we play different roles such as energy vampires, or the pundit, or the controller. Change happens when we choose to become someone else.

CHAPTER 5
EMPATHS, PARENTING, AND RAISING SENSITIVE CHILDREN

C hildren don't generally comprehend emotions. These emotions are amplified a thousand-fold for young empaths. Since they can have trouble understanding their feelings, it tends to be unfathomably tricky for empathic children to grasp that the emotions they're feeling aren't really their own. If you're an empath, you can most likely intuit whether your youngster is as well. Individuals who aren't might experience issues. They will perceive the empathic capacities in their children but can't make sense of it. There are some rules that can help you support the empath.

How Might You Tell If Your Child Is An Empath?

Most children show some level of profound clairvoyant attunement with their friends. However, some are unmistakably more empathic than others. The attributes recorded underneath are only a couple of approaches to figure out where your children's capacities lie.

Exceptionally Sensitive Or "On The Spectrum"

As a matter of first importance, they may have been analyzed as being highly sensitive. This could have been done by teachers or child psychologist. It might have even been proposed that they have real handling issues or an autism spectrum disorder. Empathic children are not only incredibly sensitive to the energies around them, they're also generally sensitive to a wide range of sensory stimuli. Many have a wide scope of nourishment sensitivities. Others may break out in hives when their skin interacts with specific textures or cleansers. Rather than simply excusing their sensibilities, it would be ideal if you attempt to respect them.

Rather than constraining them to wear a scratchy sweater that makes them go ballistic, let them pick their attire. It might make the grandparent who sewed it sad, but the child shouldn't be made to feel terrible.

If they have issues with specific nourishments, figure out what these issues are, and fix them. Do they like chips? Does it cause loose stools? You can work with that.

They Cry When Others Are Hurt Or Upset

Crying is a quality that most empaths can identify with. This characteristic appears at the earliest stages. Does your youngster cry when they see others – human or creature friends? What about when they get injured or upset? Do they hurry to comfort the individuals who are crying? Most children instinctually attempt to comfort and mitigate other people who are vexed. This attribute can either decrease or increase as they grow older. A few babies will turn out to be exceptionally self-absorbed, while others keep up their sympathetic sensitivities.

Teach your children to reflect on the damage others can cause and how they can be influenced by them. Inquire as to whether they're feeling pain or hurt feelings. On the off chance that they don't have the foggiest idea, center around what they can smell, contact, hear, taste, and see. These reflections will bring them back to the present.

When they've quieted, praise them for being thoughtful and worried for other people. Help them soothe the feelings of separation. Composing a card or a letter or prepare treats shows care and worry, without taking on the other individual's pain.

Feelings Run Deep

Empathic children regularly feel things substantially more profoundly than others do. Though one child may disregard a scolding and return to playing, an empathic child might be crushed.

Not exclusively will they feel profound hurt due to criticism. They will feel awful about disappointing a parent. Being berated by their companions will cause shame. Since they can't control their feelings, they will feel more blame and disgrace. These children are like a layered cake. They are mindful of what every other person is feeling, which amplifies their emotional reactions.

Whatever it is they're feeling at the time, they feel it more strongly than most children. Their delight is greater than their despair. Don't discredit what they're feeling, and don't ridicule them for their emotional reactions. A child who's derided or prodded when they skip or run may learn rapidly that their profoundly felt bliss can't be communicated. Same goes for their distress.

They Need A Lot of Alone Time

Much the same as grown-up empaths, children hunger for isolation. They are probably not going ever to get exhausted, because how might they be able to?

A large number of these youngsters don't simply like being separated from everyone else; they need alone time for various reasons. Calm alone time is imperative for them to energize if they have been overburdened. It's like the way skin takes time to mend after a cut. Don't criticize them for being "introverted," or request that they are progressively drawn in with others. You can't draw blood from a stone.

Adults who are depleted after nerve-racking days at work can communicate that they need quiet and isolation, and have their desires regarded. Children are fundamentally helpless before the grown-ups around them and feel like they need to give in to requests for social action or, in all likelihood, they'll be rebuffed. Regard their desire for solitude and know it has nothing to do with you. You aren't being dismissed. It's not bad that they need alone time as opposed to playing with other children. Odds are your children will welcome you for protecting their alone time.

Mysterious Physical Symptoms

Your little empath may experience the ill effects of regular stomach throbs, migraines or sore throats. Specialists probably won't discover anything wrong with them, yet that doesn't imply that the torment isn't genuine.

Regularly, these issues can emerge from the extraordinary feelings felt by the child, which will truly show if the child can't communicate to discharge those emotions.

Uneasiness or upset will gather in the belly, causing distress. Strain and disappointment may cause a serious headache.

It's extremely imperative you don't simply excuse these side effects. It can include anxiety or obsessions.

Science has shown that feeling and stress can show up as physical torment, aggravation, or endocrine interruption.

Believe your child's symptoms. Tell them that you trust them, and promise them that you'll cooperate to assist them with feeling good.

If they have a sore throat and conditions like tonsillitis and strep, have it checked on.

Do they sense that they're not being heard? It is safe to say that they are unable to talk about their reality because of a feeling of worry.

Make some popsicles. They will help a child feel happy so they can talk. Assist them with communicating through composition or waiting until they're ready to verbalize.

Do they have stomach aches? That is typically identified with stress or anxiety. Peppermint tea or ginger can be useful, and afterwards, non-critical profound stomach breathing and gentle yoga.

Once quieted, check whether you can work with them to make sense of where the irritated is originating from, and check whether you can discover solutions together.

WHAT TO DO IF YOUR CHILD IS AN EMPATH

The Child Empath

As children, we are generally empathetic. As children, we are on the whole profoundly sensitive to nature and empathic to life around us. The issue for all of us is that, as we grow up, society thrashes us, and we become (unfortunately and pointlessly) detached, desensitized and shut down accordingly. A few people, nonetheless, can't be closed down. A few people in these present circumstances are empaths, and these individuals don't lose their association regardless of how much poisonous brutality and mistreatment they experience. Staying empathic "regardless" might appear as though it is something to be thankful for. However, things aren't always rosy. Our reality is very fierce and poisonous. People who are empathic adapt to being powerless. A lot of the time, their family exposes them to negative energy. This can also come from companions, co-workers and strangers. They cope by finding a way to deal with the situation. Much of the time they still feel devastated. Indeed, even the most grounded empaths need to be reassured.

The question before us today is, how would you appropriately secure and bolster an empath? It is an important question. Right now, I would prefer not to talk about grown-up empaths. That is a significant theme that requires broadened treatment. Instead, right now, we need to give a touch of direction to guardians and instructors on the best ways to prepare and ensure their empathic youngsters get help. Before I get into that I need to state that the guidance I'll be offering here is not entirely the same as the guidance empaths (and those related with them) usually get. Ordinarily, empaths and those related to them, are given what I would call "independently engaged" exhortations. The independently engaged persuasion is guidance coordinated only by the empath. When empaths, and the individuals related to them, get this kind of appeal, they are advised that it is up to the empath to adapt! Empaths are informed that on the off chance that they need to "bargain" with their endowments, they will need to figure out how to "adapt" with the real world. To adapt, they should "reframe" their viewpoint, clean their home, cry and express, use precious stones to purge energy, manufacture new limits, and perceive they're perfect crucial, and so on. I'm sure on the off chance that you've, at any point, gone over the "standard" appeal given to empaths, you'll be comfortable with the direction given. Shockingly, I need to state, exclusively engaged adapting systems genuinely don't help the empath because the issue isn't the empath. The question is how they should persevere in bad conditions. Independently involved changing systems are exceptionally hazardous because they either disregard the hugeness of the situation's effect on the empath or acknowledge the earth wherein the empath moves as an unchangeable, even profoundly alluring given. These methodologies take the brutality and lethality of this world with a "well, that is simply life, better get used to it" disposition. A few defenders of exclusively engaged systems even venture to state that all the poisonous quality and viciousness that empaths experience is really a profound "test" or "exercise". They state to the enduring empath, "what doesn't kill you makes you stronger," and they leave it at that.

Empaths are here as healers to help the individuals of this planet recuperate and push ahead. Empaths are particularly sensitive and "emissaries of the Consciousness or Spirit," and it isn't their business to adapt and acknowledge it. Despite what might be expected, we must shield them from brutality and mistreatment so we can ensure that they remain sound. They need to maintain their goals so they can grow up and carry out the responsibility of adulthood.

If an empath falls under the unhealthy weight of this planet, it is our fault, not theirs.

Conditions helpful for sound sympathy require fundamental worldwide change. This is everyone's responsibility. That doesn't imply that we need to trust that global change will begin fixing things. There are things that you can do beginning right now to improve conditions for the highly sensitive empaths in your lives. One of the most significant, huge and perfectly possible things we would all be able to do is to protect all the offspring of this world. All people are empathic in adolescence, and some empathic are exceptionally open and associated. If we need our youngsters to grow up empathic, and If we need our empaths to grow up sound, entire, engaged and prepared to recuperate, we need to begin by helping them in childhood. I am sure you will concur. It is during the long stretches of our childhood and puberty when we are generally helpless against the stress of adults in the long periods of adolescence and immaturity, and we are most needing assurance. Given that, here are a few things that guardians, educators, scout pioneers and other people who are day by day associated with children can do to begin making living conditions. That is sound, strong, defensive, and helpful for the advancement of completely working empathic healers.

Parents and teachers

Perceive that empaths exist: The principal thing parents and teachers can do to help is to understand that empaths exist. This is a significant initial step. If you don't accept that empaths exist, you won't put any psychological vitality into seeing them. You must bolster them. We must figure out as a real need how to see empaths ; otherwise, we'll get not even close to making a difference. Along these lines, the initial step is just to accept that empaths exist and hence put some psychological exertion into seeing them in our general surroundings.

Comprehend empaths: Once guardians and educators have perceived that empaths exist the following stage is to endeavor to understand them. Fortunately, that isn't advanced science. To comprehend empaths, essentially see how they act and interface with others and how others act and communicate with them. Our motivations here are that they are exceptionally sensitive to physical and non-physical, immediate and circuitous, dynamic and aloof attack. Assaults of various types hurt an empath. Shouting, hollering, boisterous attack, verbally abusing, jokes, prohibition, egotistical dismissal, disparaging hits; and even just arbitrarily diffused, non-coordinated energies of outrage, threatening vibe, misery, and depression can hurt the empath! As noted, numerous individuals desensitize themselves so they are never again able to feel the agony. Empaths can't desensitize and so are continually helpless and subjected to attack. This is a significant thing that you should always remember. Regardless of how solid you figure somebody is, attack them enough, trouble them with enough toxic mental and passionate muck, and they will breakdown under the strain.

Perceive Reality: Speaking about toxic muck, the third thing guardians and educators can do to support the empathic youngster (or grown-up) is to perceive the genuine idea of our world on this planet. This is significant. The picture that Hollywood presents to us is bogus. We don't live on television shows. Our families are broken, our general public is brutal, and our social connections are conspicuously shallow and frequently poisonous. Children on this planet are misled, beaten, mishandled, abused and assaulted every second of their lives. Indeed, even honest "sleep overs" or "scout outings" can open our children up to danger, mistreatment and rape. It is not just child abuse, child labor, children as erotic entertainment, that is the problem. This planet is a harmful place for youngsters, and that is the truth. Denying it, imagining it isn't there, believing there is advantage in brutality and mistreatment, or essentially overlooking it won't keep children from being mishandled and empaths from being hurt. If you need to help the empath, open your eyes and see.

Perceive that harm occurs: Once you see and comprehend the empath, and once you are on top of the real world and not in a dream, the following thing is to understand is that harm does happen to the empath. If you hit an empath, if you shout at them, if you control them, If you mistreat their energy, if you attack them, if you belittle them, if you assault them, if you attempt to muddle them, if you absorb them, harm occurs. It hurts like being cut with a knife. You can imagine that isn't the situation or perhaps mollify your complicity and blame others by making statements like "what doesn't hurt you makes you stronger" or "they decide to be here" or some other nonsense.. However, if you do, you're deluding yourself. Attack a youngster, assault them, and the empath will feel beaten. You are harming them and causing them harm, period.

Comprehend that after harm it takes a long time to recuperate. Once you comprehend this and understand the empath and the danger they are in, then you can comprehend that passionate, mental and profound harm occurs. This is so because of toxic socialization. It will take a very long time to mend. Physical harm can take as little as a couple of days to recuperate. But researchers and others are starting to acknowledge enthusiastic, mental, and otherworldly harm brought about by poisonous homes and school conditions can take a long time to mend. The financial expense to society alone is not being considered. It doesn't take a lot to harm an empath. Indeed, even a solitary enthusiastic, or a mental attack can do harm. It is a genuine social and mental issue. We need our empathic children. and all children, to grow up solid and entire. We have to begin focusing on the genuine outcomes at this moment.

Issue a NO ABUSE decree: Once you perceive that empaths are inconsistent peril, when you perceive the high, long haul cost and the genuine need to secure our youngsters, the following thing that you should do is STOP THE VIOLENCE any place you can. Issue a "NO ABUSE" decree. In our home, we have zero resistance for maltreatment of any sort. In our home, no one is permitted to shout, call names, yell insults, take part in dynamic or detached attacks toward someone else. This doesn't imply that we are not basic or open, however it means that we attempt to impart in a positive way consistently. When enthusiastic attacks do occur, the culprit is responsible and should apologize. It is often the grown-ups who are the culprits and the youngsters who are the victims. Contrasted this with the adolescence of numerous children. The home needs to be safe. We need our youngsters to grow up sound, entire, associated and empathic, and this won't occur if we attack them.

The equivalent "no treatment" rule ought to be received by

educators in schools. Instructors are mistreating children, children mistreating one another, and any type of mental, passionate, or otherworldly viciousness should stop. There needs to be serious consequences in the class or on the schoolyard when this happens. Attack of any sort harms the empath doubly so. There is no space for maltreatment in the life of the youngster. Issue a "no abuse" proclamation.

Secure your children: Issuing a "no abuse" announcement. It is the main advance that you can take. When you have announced your decision and want to live in a maltreatment free condition, push for its help, appropriation, and requirement. What's more the house isn't the main spot that youngsters and empaths endure attacks. Mistreatment could occur at school, at grandmother's, at aunt's, at playgrounds and amongst friends. You can't control where a grown-up goes, but you can shield children from harm. Never open empathic children to danger. Keep youngsters far, far away from dangerous grown-ups and harmful children. Playdates are innocuous when guardians are dangerous. Poisonous qualities seep into youngsters. They can start to carry on in harmful ways towards other children. In the long run, other children become the culprits of viciousness. Every single creature is equal. As empathic children develop, they become increasingly presented with and vulnerable to antagonism. The initial years when youngsters play together is not the difficulty. However, by ten, twelve, or older, the danger and mistreatment that can be experienced in families begins to grow. Hence, shield your youngsters, teenagers, and young adults from others. You may believe that guiding your children into family and social circumstances is a positive thing. However, it is frequently not. With regards to securing teenage empaths, adopt a no-bargain strategy.

Encourage your children: Without a doubt, guardians, and instructors must take care of youngsters as they can't look after themselves. Youngsters grow up to be adults. Guardians and instructors should release control. At the point when youngsters become grown-ups, it is dependent upon them to ensure themselves. They can do this if they have been instructed what to do. They don't realize what to do by magic! Guardians and instructors need to show youngsters how to secure themselves, how to adapt to cynicism, how to construct limits, how to dismiss mistreatment, and how to avoid risks. One of the most significant things that you can encourage your child to do is explain to them that there is no excuse for mistreatment. In the event that you instruct them that there is no excuse for mistreatment, at that point they will grow up dodging it, dismissing it and declining to acknowledge it. If you instruct them that mistreatment is alright just like when they are disciplined, they will be less likely to remove themselves from injurious circumstances. Keep in mind that harm levels increase when empaths, children, and even grown-ups experience mistreatment. Educate them that there will never be any legitimization for mistreatment. If mistreatment occurs, they must realize they should leave. This is particularly significant for empaths. Mistreatment harms everyone; children and empaths doubly so. In this way, empaths must be educated from the very beginning that no measure of physical, enthusiastic, sexual, mental, or otherworldly maltreatment is adequate.

Instructing children to dismiss all types of mistreatment may sound basic and direct: numerous individuals will battle powerfully against mistreatment. Think about this: the essential message that guardians and educators send to youngsters is that it is alright to hurt another human if they deserve it. The message that children take away if someone accomplishes something awful, is that one should feel free to hurt them. We may not talk plainly with our youngsters. We show them practically day by day. Guardians, educators, and grown-ups appear to experience no difficulty when it comes to shouting, hollering, hitting, harming, disgracing, putting down and in any situation attacking the unprotected child when they do something incorrectly. We attack our children and afterwards to legitimize it, we disclose to them that they deserve it. That is only the beginning. As children grow up, they hear the message and see it demonstrated over and over again. When we are completely developed, we can legitimize significant degrees of mistreatment, even torment, by just revealing to ourselves that the individuals we are harming deserve to be harmed because they did something awful. The message is completely clear. Our children learn we can hurt them.

In any case, that thought isn't wrong. That message needs to change and it needs to change now. We, as a people, as a country, and as a planet, need to know there is no reason for mistreatment. Regardless of what the youngster, youthful, or grown-up does, mistreatment is never a choice. Discipline is abuse. Individuals are once in a while "rectified" or "transformed" by abuse. Generally, they are simply harmed much more. There is no advantage. Attack harms like damnation, and it causes a weakening to the body and mind. This is particularly so when we think about our empathic children. Remember, all children are normally empathic. We need them to grow up sound, entire and solid; we need to stop hitting, harming, shouting, and beating them. You can help with that by halting the maltreatment and sending a reasonable verbal message to the children and teenagers that mistreatment isn't OK. Try not to show children they should just put up with it. Show youngsters to recognize it, keep away from it when they see it, report it to individuals who can take care of them, and oppose it.

Invest loads of positive energy: The world will never be an ideal place. In any case, even in a perfect world, there will be times when attacks will happen. We're all in a human body and the human body loses power. In any case, that is OK since people, empaths included, are flexible. Empaths are amazingly sensitive; they are not feeble or gutless. They need to be instructed to build safe spaces and limits. When they have been educated to dismiss mistreatment, they are solid and strong. We can manage the aftermath of someone's loss of control because the physical body is strong. It has worked in adapting and mending systems and these tools, if they are working and appropriately "charged", can manage injury. You can help by basically giving them affection. At home, this means loads of cuddles, love, consideration and support. At school, this means lots of positive respect, consideration and support. Empathic youngsters, and all children, need lots of consideration, loads of adoration, heaps of suitable physical contact, and heaps of help. More than they at present get. As a parent, we will be facing time constraints in our day. Instructors will face class sizes and regulatory rights. My advice to guardians is to organize your children's' enthusiastic and mental needs over drinks with the guys, mother's espresso club, etc. My recommendation to educators: train yourself to focus and push your colleagues and administration to stop focusing on death, like war. Begin loving the earth. We are rich social subjects, and we can manage the cost of smaller class sizes. Mechanically and financially, we have all that we need to make the ideal world. Reject governments that are by the rich, for the rich. Choose governments that believe individuals and do things for them. As guardians and educators, we are the ones that must choose governments that will help our children and teenagers.

7 steps sensitive mamas can take to discipline sensitive children

Mothers feel the pressure rise as their morning moves through stormy waters. More often than not, the peaceful stream is carrying us along, yet right now, there is a tempest circling overhead. The eye has all the earmarks of being slowed down over the family room where the little youngsters are contending.

Thankfully, as a rule, children are usually the best of companions. They are kind, cautious and innovative. Like all imperfect creatures, they falter now and again. I realize this is advantageous in building up the aptitudes to coordinate and bargain in your life. Yet it isn't enjoyable for the middle person, particularly one that is sensitive to everything confusing.

As an individual who flourishes in calm, upheaval can be nervous. For some sensitive individuals, quiet is restoring. But even with excellent child-rearing, there are constrained snapshots of stillness. Peaceful time must be deliberately made.

It is the individual's mission to comprehend themselves better and parent in my most calm manner. They should read about character theories and empaths. Workshops about delicate individuals are helpful. Parents can even speak with different guardians. Take time to think about it until you come to an understanding.

Permit yourself to consider that you may be the parent of a sensitive individual. You need to serve them and teach them manners. Managing children as an irritable individual can be a test, particularly with regards to training.

1. Teach

Embrace the conviction that parents need to instruct, not rebuff. Order originates from discipulus, the Latin word for student. That is one who has been instructed or affected. Instructing is a lot gentler than rebuffing, for both the instructor and the student. Children want to give, so we must show our youngsters how to treat others with thoughtfulness through example. They will realize how to live.

We can decide to find the reason for conduct as clarification. We can consider ourselves to be defective and offer help because we know how it feels to be mistreated. We can be the change in a child's life. We can give them a great deal of direction.

2. Pause

Decide to delay and loosen up before responding. Take a full breath or step away for a minute. We should be quiet to make quiet. A full breath has the ability to loosen the body. If we have a calm psyche, then we can permit access to our most tranquil method for being. This is the place our most noteworthy direction begins.

3. Feel rather than fix

Most guardians need to shield youngsters from anything upsetting. It is helpful to permit uneasiness to outfit children with the tools to overcome anxiety. Permit space to fortify your relationship. If we give, all children will learn is to take. Strengthening comes when a child feels support and wants to investigate.

4. Express feelings

Offer yourself the option of letting your youngster show their feelings. At the point when we permit feelings to be communicated, we show our youngsters that it is alright to feel upset. They can come to understand they will discover what it means to be quiet. "I see that you are disappointed, I feel that, as well," is an incredible expression that makes a connection. With affirmations, we can travel towards harmony.

5. Make it matter

A child can't process feelings autonomously. It isn't good for anyone to be a sitting time bomb. Sit together and quiet your youngster. Afterward, correct the conduct through delicate direction. Your consideration shouldn't be seen as a gift; it is what is expected of you.

6. Empower connection

Embrace connection because they need it as much as you. Children will participate as much as they feel like we care. A sentiment concedes goals, while absolution is the result of that.

7. Organize self-care

Realize that self-care starts as the mentality of accepting that you offer more when you are rested. We want our children to live their best lives, so it is significant to show them, with our decisions, how to adjust. There is a period for service and a period for stillness. Both are basic to being our best selves.

How might I comprehend my child's perspective better?

Being an empath implies you have super-charged sensation capacities. You FEEL feelings considerably more and have an elevated sense of vitality. It's a blessing that you can rapidly comprehend inconspicuous hints. If your youngster is empathic, they know when you are concentrating and miserable and when they need to share or sympathize with you. They will recuperate you with a hug. They can take the worry away from you. Empathic youngsters are grounded in mindfulness. Consequently, they can be pushed to the edge more quickly than other children. If you consider all the difficulties individuals face at school and in everyday life, you will also see worries radiating like radio waves. Your empathic child resembles someone absorbing feelings and holding on to them. Generally, the outcome is a physical affliction like nervousness, misery or a feeling of being overwhelmed, a tight chest, asthma, bronchitis and diet hypersensitivities.

Top 4 things a parent can do to support a restless youngster.

Focusing on children can be difficult if you don't understand why they are feeling on edge. Realizing that your little one is merely responding to their general surroundings, including what is going on in the family, it is important to know before fixing the problem. Help your child to quiet down and feel more joyful in their everyday life with these steps.

1. Earthing routinely

Earthing is the basic physical procedure of tying down the human body to the earth. This makes them feel heavier and more present to their bodies. This abates uneasiness in youngsters and thus relieve feelings of anxiety. Go outside into nature or to a recreation center, guarantee your youngster has their shoes off, and get them to go through 15 minutes of lying, running and playing on the earth. It is PROVEN to unwind and de-stress individuals because the planet has electromagnetic vitality that beats through our feet and reconnects them to their natural energy.

2. Energy Clearing

If you recall that feelings resemble radio waves and that pressure can get stuck, then you can perceive why getting the vitality out is significant for empathic youngsters. I like to utilize white sage. It is a herb to consume and smoke. It is used for cleaning the room. At and the same time, set positive expectations, like "I welcome upbeat, positive emotions into this house and room" or "I let go of pressure and weight." It's stunning the impacts this has on the mind of the house or individual. You can also attempt the strainer system. Request that your youngster picture a sifter. Without brushing their body, the entirety of the mess flows out of their body as it goes up and out over their head. Start at the feet when doing this exercise. Do it with them at regular intervals or when they feel angry. Do it twice because each subsequent time more pieces are caught in the net.

3. Breathing and Meditation

We realize that breathing is the component of the heart chakra (energy) point. It has been demonstrated by science that breathing deeply has a quieting impact on the body. Take a stab at consolidating ten full breaths into your day to help make a quieter child. Request that they picture taking in green light. The color green is known for diminishing anger and acquires harmony; it is the color of nature!

4. Creative activities

Uneasiness and stress are a left mind thing. The left side

represents rationale, mental and sane states of being. Utilizing this piece of the mind creates pressure. If we switch to the opposite side of the brain, the right side, we enter the region of innovativeness, internal identity, and instinct. This is a tremendous pressure discharge for children. Empaths particularly need to utilize their hands, so try planting, cooking, painting, or dog training. Animals are superb for empathic children since they are so genuinely cheerful and create great vibes.

CHAPTER 6
EMPATHS AND WORK

- Being an empath means that you assume the feelings and pains of others — which can permit you to be both incredibly conscious and have a more grounded response to pressure at work.

- Just retaining this pressure can be a risk to your well-being.

- Instead, set aside the effort to inhale, thoroughly consider the realities of the circumstances, use a peaceful area to ground yourself with what you're grateful for.

If you're an empath, you can experience a tremendous measure of compassion. It gets to the point of assuming the feelings and pain of others at your own cost.

Do you cry at commercials? Do you quickly observe everybody's side of the story? Do you consider yourself to be the characters in books and films?

As an empath, stressful, testing or turbulent circumstances in your work life can feel overwhelmingly stressful. It gets to the point that you need to take a day off to recover.

Absorbing these problematic circumstances isn't serving you — it's a threat to your well-being. But avoiding these times is impractical. Each activity has its arrangement of good and bad times, and it costs you your valuable PTO.

Anyway, what are we empaths to do? These emotions are essentially signals for you. They say it is a great chance to recalibrate. To manage intense work circumstances as an empath, we start with our bodies and then deal with our psyche. Here are five different ways to manage difficult work circumstances as an empath.

1. Relax

To delay the circuit of negative emotions coursing through the body, retreat to your vehicle or the restroom and slow down to relax. You can also do this activity in your work area for 30 seconds.

Close your eyes. With your hand on your gut, feel your hand raise as you inhale air in through your nose. At the point when you feel full, hold your breath in and count to three. At that point, gently push the air through your mouth. Remind yourself with every breath: "This is my body. This is my breath. This is my mind. This is my minute."

2. Name feelings

What do you feel precisely? Naming your feelings is essential, but it can take some training. Schools teach that the ideas of "happy" and "sad" are the only emotions we're equipped to experience. There are more.

Whenever you have the feeling that you're over-sympathizing, pick a feeling word to depict what you're experiencing. Search online for a list of emotions if it's difficult to name one that feels spot-on. At that point, state to yourself: "I am feeling the feeling of ___."

Note where in your body you are experiencing that sensation. Tension is felt on your shoulders or it feels like your chest is caving in. Naming works for positive feelings, as well! Bliss may feel like effervescence in your sternum or warmth in your cheeks.

Express your tactile feelings truthfully. You will turn into a more peaceful observer of the whole situation.

3. I figure out what I am thinking and whether it is serving me.

Ask yourself this series of questions to dial into the thoughts that are driving your emotions:

1. What is my assessment of what's happening?
2. Why does this make a difference to me?
3. Why do I need this outcome?

The responses to these questions decide what you will think. These musings are driving your emotional experience. Presently, consider: What are the provable facts of the circumstance? Take out all feeling. Using this list of realities will help you to be able to pick what you need to make the condition meaningful.

It is the empath's choice of how to interpret circumstances. It is food to order and practice parsing out the facts from the musings. It takes intense thought.

4. Make a "quiet zone" in your home.

A "calm zone" is a no telephone and email zone found in your home. Leave your telephone on airplane mode and in your purse, in your vehicle, or in the pantry. You can protect yourself from the impulses and feelings of others by killing that channel during your alone time.

5. Offer thanks.

It's the best cure for anxiety. It doesn't need to be anything momentous. It doesn't need to be a family or friend's thing. It's better if it's not. Try a clean drinking water, fluffy socks or scented flowers. The sound of a good house.

If appreciation feels like something you need to implement, consider beginning a notebook. Put it next to your bed or your bath. Somewhere you will see it regularly. And dare yourself to write down one thing you're thankful for. Do this consistently for seven days. Next time you feel empathy overpowering you, picture that notebook's list of things you're grateful for and channel your feelings into that.

Be sympathetic to yourself

As an empath, feeling the emotions that others are experiencing is how your mind is wired for survival. Your mind has been working superbly. Being an empath is an advantage and can serve you well in numerous ways. You can peruse a room or meet individuals where they are. They feel like you genuinely get them, and you know who's disclosing relevant information versus those who are hiding something. You get vibes off of individuals with laser-like accuracy.

The most effective method to Thrive at Work as an Empath

Half a month back, I discovered that my best friend had lost a friend or family member. I was working two jobs in an office at that point, and immediately unraveled at work. I sensed that I was going to cry and decided to remove myself to a nearby park. There I could cry in peace.

That wasn't the first occasion that I'd felt a shockwave of

emotions at work. I'm not unprofessional. I'm an empath. Empaths are exceptionally sensitive to the feelings and states of mind of others. In contrast to sympathy, in which you can understand what another person feels, empaths will, in general, feel more.

Empaths fundamentally don't have the same channels that others have, so they feel everything. Ask yourself if you have been called excessively sensitive? Do you retain the feelings or physical side effects of others into your own body? Do you tend to read between the lines?

This degree of affectability can regularly be seen as a shortcoming or absence of flexibility. Dealing with your empathic characteristics, you can work to further your potential.

Instead of being critical, try to feel what others are feeling. Research recommends that sympathy and empathy in the working environment can expand profitability and strengthen employees in tough times.

And, it can deliver insight into your friends, bosses and workers in the same way. The emotions and feelings you can gather from others are only one option that gives us more to work with. You get more information, are able to pick how you will see things going ahead, and decide what you will do with that data.

Therapeutic services and directing jobs require good

observation skills. It is a result of sympathy. Emotional well-being is critical if you want to achieve great work.

A significant piece of the puzzle to effectively balance an empathic character in your vocation or a position of authority is figuring out how to manage them. You can cry in an open park a limited number of times before individuals begin to ask where you are. Here's how to take advantage of it.

Approach Your Feelings Objectively

Like a system utilized in cognitive behavioral therapy, identify your feelings that show up when things get tough. Envision yourself as an artist observing your thoughts and think about what you need to find peace.

At that point, remember that you're still in charge. You're the artist right now, and you get to decide which thoughts you're going to let matter and how much. Removing yourself from passionate circumstances makes that simpler. Another approach to consider has to do with reflection, in which you recognize your thoughts or emotions and then let them go.

Control Your Environment

Since empaths resemble sponges for feelings, it's useful not to have any sentiments to absorb. Certain office conditions, like those with an open floor plan, can be challenging to explore since there are no limits or desk dividers.

It also assists with taking stock of your remaining burden and timetable all the time. "What number of customers am I finding in a given week? How late am I working? What kinds of cases am I taking on? With such a large number of customers, late evenings or tough cases with no rest can quickly become draining.

Visualize Your Protection

Regardless of whether someone you see at the workplace overlooks your "Hello" or an associate undermines your thoughts during a meeting, negativity can be infectious—and particularly so for empaths. It could easily cause your state of mind to take a crash, which then may influence efficiency and execution. It can remain long after you leave the workplace.

This nonexistent boundary can protect you from coincidentally absorbing tragic, furious, bothered or negative emotions from another person.

Take a Three-Minute Meditation

At the point when sentiments have overpowered you, give yourself a break.

This is especially useful when you're responding to positive or surprising news in the working environment. It can be overpowering to anybody—empath or not. For the most part, you need to get yourself away from the gathering. Even if you can only get to your vehicle, it's better than melting down in the workplace.

Think of it as Instant Feedback

Let's assume you're presenting a plan to a customer or a boss and get the feeling that your audience isn't into it. Rather than letting it discourage you or causing you to think of losing control, use it as a sign you need to recalibrate. If anything, the capacity to read the room rapidly is a benefit. The feedback causes you to target your information so you can more readily and clarify the fact of the matter that you're attempting to make.

CHAPTER 7
EMPATHS, INTUITION, AND EXTRAORDINARY PERCEPTIONS

A natural empath is highly sensitive to the feelings of others and is ordinarily a very good listener. An intuitive empath will regularly pick up on someone's feelings even if they are covered up. The exceptionally delicate nature of the instinctive empath can help heal others. This is especially true for those who have a history of emotional or physical misery. This is one reason natural empaths have become well-known types of holistic healers.

Figuring out how to understand and process your feelings. Certain college courses will help you turn into a healthier and happier person.

What is the difference between Intuition and Empathy?

Instinct is characterized as the capacity to see reality without having to think about it. Instinct can show itself in a variety of ways. Sometimes it doesn't require any special understanding or conscious effort. People have various degrees of instinct, and you can take a shot at improving your own instinct with courses. Compassion is characterized as either mentally or vicariously experiencing the contemplations, emotions or perspectives of others. Compassion is unique in relation to sympathy. With sympathy, you truly feel someone else's emotions, while compassion is a feeling of caring about someone or something. For example, if you are frustrated about what someone else is experiencing. If a person can see the feelings of others, particularly the deep, hidden feelings, that person would be marked as a natural empath.

Natural Empath Healing

In holistic healing, emotions assume a critical function. Negative beneficial experiences and feelings can make an imbalance in the chakras, or life energy. At the point when energy gets out of balance, it can show as physical illness. So as to heal totally from the physical infection, an individual needs to figure out how to mend the energy imbalance. Acupuncture, Reiki, and chakra healing are only three instances of healing models that work directly on healing the bad energy bad in the body.

An intuitive empath can encourage the acknowledgment of the negative feelings or energy that adds to the illness. Numerous individuals suppress feelings and negative occasions. When that happens in their lives it can be caused by injury or catastrophe. These feelings can be covered up to such an extent that even the individual doesn't remember they exist. In any case, they can even now cause physical and mental issues. A natural empath can perceive these feelings and emotions and expose them, allowing the individual to process them, for example, through the appropriate treatment. This can help to rebalance the life energy in the chakras and take an individual back to full well-being.

Step by step instructions to know whether you are an Intuitive Empath

Empaths are regularly talented onlookers and can understand things by an individual's non-verbal communication and the way they move. Most empaths are oversensitive. They feel overpowered in large groups, so they like to be distant from everyone else or with only a couple of others. Empaths regularly need to revive and relax. This means being all alone. Empaths are regularly magnanimous and will try to help other people. This could hurt them. They also can see directly through untruths and perceive the motivations and intentions of others. Empaths also are frequently sensitive to surrounding noises and sounds, including others talking. At last, empaths frequently will experience serious emotional episodes that are not on top of what is happening in their life. Many empaths are imaginative and become specialists, performers or writers. Instinct can be a significant component of being an empath.

The Dangers of Being an Empath

Numerous empaths feel the feelings and emotions of others truly, as opposed to simply inwardly. In this manner, one of the threats of being an empath is assuming the feelings and emotions of others. At the point when others feel glad and positive, this lifts the empaths own feelings of happiness and joy. Many around them are loaded up with negative feelings. The empath can find oneself loaded up with negativity and even fall into a deep depression. These feelings can add to sickness and medical issues in the empath.

Types of Empaths

There are two essential kinds of empaths: emotional and cognitive. An emotional empath genuinely feels the feeling and energy of others, which allows a greater understanding of the feelings of others. Some people don't know about their specific feelings. A psychological empath wants to understand a person by placing themselves into the other individual's shoes.

There are subtypes of each of these empaths. A manifested empath can make their world and implant the truth in others. A natural empath can detect and feel the feelings of others. A healing empath can assimilate the energy of others to encourage mending. A profound empath can detect how the universe comes together. A precognitive empath can anticipate different futures. A telepathic empath can recognize thoughts just as sentiments. A judgment empath can recognize truth and lies. A seer empath will have the option to understand the reason for feelings, however where they show. There are rarer types of empaths too: an atomic empath that blends auras, an animal empath that speaks with creatures, a shaman empath that cooperatives with nature and a general empath that has all the characteristics.

CHAPTER 8
THE GIFT OF BEING AN EMPATH

You have an innate capacity to see everyone around you. In any case, being an empath is something other than being a particularly sensitive individual, and it isn't restricted to just feeling and expressing emotions.

It is frequently said that empaths are normally natural, and keeping in mind that this is valid for some empaths, it isn't adequate for all. But, if somebody is intuitive, it doesn't imply that they have a great deal of sympathy. Some accept that these words are tradable. However, the two capacities are quite different.

Empathy is one's anxiety with things other than oneself — the outside world. It is the capacity to detect the emotions and energy of others and one's environment.

However, intuition is the internal emotions one needs to assess and comprehend a circumstance. While everything includes absorbing the outside world, the last part happens inside your mind.

At the point when sympathy and instinct meet, an

extremely novel capacity is conceived, and the individuals who hold this quality have a special gift.

What is a true empath?

A true empath is one who can feel the pain of others as well as their own. They can also feel the delight that others feel.

Along these lines, a true empath frequently prefers to help individuals make everyone around them happy and accordingly have the option to feel that happiness themselves. Empaths are great listeners who never overlook individuals and don't wish to be false or put on a veneer.

These characteristics are stable. They could apply to a lot of good individuals, yet a genuine empath encourages others to the point of fatigue. They are also sensitive, not exclusively to individuals' feelings, but in addition to external factors in the environment, for example, light, clamor, and movement.

Natural Empath Traits

There are numerous attributes that instinctive empaths share. For instance, they strongly dislike seeing other people suffer and frequently abstain from viewing the news or even sad movies for the same reasons.

Subsequent since they feel strong negative feelings, natural empaths are then prone to feel physical symptoms in addition to their emotional despair, for example, a headache or fatigue.

Instinctive empaths tend to become attached to individuals

who are out of luck, regardless of whether it is or isn't in their own interest.

This might be a stranger or it could be a partner. This puts instinctive empaths in risk of being in toxic relationships.

A number of these qualities may seem negative, however being an empath is a blessing from various perspectives. Here are 17 reasons why this ability to understand others is really something to be thankful for.

Reasons Why Intuitive Empaths Have A Gift

1. Empaths encircle themselves with positive things

At the point when an empath sees something bad, they intuitively dismiss it so they don't feel the negativity. In this way, they attempt to surround themselves with only positive things.

This leads to empaths having positive friends who want to carry on with a happy and satisfying life.

Happiness is contagious, so one of the best approaches to discovering satisfaction is to surround yourself with individuals who can make their own joy and offer it to others.

2. Empaths normally have meaningful careers

Empathic individuals can frequently be discovered in jobs such as tutors, life mentors and educators. On account of their empathy for nature, they frequently either do volunteer work for the environment or make a profession out of environmental activism.

Because of their intense faculties, empaths love to spend countless hours experiencing nature and getting a charge out of the melodies of winged creatures, hints of the sea and the smell of flowers.

Having an important vocation or offering back to society is a satisfying method to have a happy life.

3. They unite others

Empathic individuals find it deeply uncomfortable to watch others be mean or hostile towards one another.

Due to their peaceful and kind nature, empaths frequently go about acting as a middleman. They bring amicability between people who are at odds.

They have the gift of being peacemakers and decrease the stress of others by decreasing any disdain that is being held.

4. They don't race into deciding

They look for a deep authenticity before deciding. Empaths think that it's troublesome to settle with an answer if they feel it might have been negatively impacted.

Since they like to remain consistent with themselves, they will set aside the essential effort to come to informed decisions.

This is a blessing because they are able to postpone settling on time, to allow people more choices. They don't like having regret in the long run.

5. They can experience the feelings of their friends and family, regardless of whether they are not with that individual

It is normal when an empath is in a strong relationship with somebody that they can experience unexpected floods of feeling or pain. Later they can discover that their loved one was feeling similar feelings or pains at the same time.

Regularly, empaths describe this feeling as being "blindsided" by feelings when they are not expecting them.

6. They like to remain centered

While numerous individuals may accept that being able to perform multiple tasks is a quality for everyday life, it is generally false because none of the tasks are finished on time.

We are performing multiple tasks that partition one's attention into a wide range of things. This is done quickly leading to confusion and feelings of being scattered.

Natural empaths feel better and do more effective work when they carry out their responsibilities, each in turn and in a specific way.

Focusing on each thing in turn and doing it as well as could be expected. It can be an empath's source of personal energy that leads to great success.

7. They have a talent for tuning in to people's stories

One distinctive attribute of this populace is that they offer a dependable and loving feeling of friendship to others.

Others incline toward them and feel great trust with empaths. They talk to them about their own personal struggles. Even strangers may move toward an empath and open up to them about their issues.

This is acceptable on the grounds that empaths realize that others feel assuaged when they are finished talking about their feelings.

Empaths would then be able to feel this good feeling themselves since they have helped somebody who is remembering some bad times.

8. Empaths rush to spot somebody who is being dishonest

Individuals who are empathic can detect a deceitful person.

They pick up on the subtle hints of deception, which allows them to trust the individuals who they decide to be near because they realize the people surrounding them are not deceitful people.

Apart from this, this also means empaths know when somebody is concealing something when they state they are "fine".

Empaths know when these individuals are really crying inside and setting up a false facade.

9. They pick up on non-verbal cues

Probably the best blessing that an empath has is the capacity to read others. This allows them to rapidly choose if somebody they meet will be their friend for life.

In addition to non-verbal cues, empaths can pick up even slight signs of others' physical needs and feelings.

This gives them a particular ability to pick out the needs of individuals who can't talk, are similar to animals, babies or the human body.

10. They can see the big picture

Having a feeling of synchronicity allows empathic individuals to get a handle on the bad plan of things since they can perceive the interconnectedness of every individual and living organism in the world.

Having the option to see the big picture allows people to feel significant in what they do and progress in the direction of their goals. These end up being more fulfilling.

11. Empaths have great imaginations

At the point when an empath winds up stuck in a mundane everyday schedule, they will float off into a daydream.

If their environmental factors are not giving any emotional stimuli, empaths can lose interest for what is happening around them and get lost in their world of imagination and creativity.

12. They are creative and artistic

Empaths can pass on a message in ways others can't. They can utilize their feelings and project their artistic and creative talents.

This could turn out in numerous mediums, for example, moving, painting, writing verse or playing music.

Empaths have a special ability for creativity with art. Additionally, their experiences, circumstances and different conditions also require their gifts.

Since empaths think uniquely, they can see certain things that others can't quite conceptualize as easily. This thoughtful creativity and capacity to process data distinctly is a notable capacity.

13. Empaths can see everyone's point of view

One reason that empaths are such good friends to others is because they are eager to tune in to and understand everyone's point of view.

This also causes them to have the option of being long-lasting students as they experience a wide range of people to learn from.

14. They are natural healers

Empaths are normal healers and can really give their healing energy to others through their senses.

They have a healing energy that can help people around them and help themselves too.

15. They have enormous energy which lasts forever

Since empaths feel everything with such quality, they are inclined to feel more prominent highs than others.

In this manner, most empaths are energetic about existence, and they can experience joy with a greater intensity, which drives them to be more kind, mindful and caring towards others.

16. Empaths are comfortable being alone

Many individuals who are not empathic feel uncomfortable being distant from everyone else. Not empaths. They long for time alone and really need it in order to adjust themselves and de-stress.

They appreciate this opportunity to recover and build their mindfulness along these lines.

17. They can add to others' lives as nobody else can

Since empaths really care about others, they can affect the lives of individuals who feel like they have nobody to go to.

This means they can have a colossal effect on individuals'

lives and leave a positive effect on individuals who are suffering. Being intuitively empathic appears to be depleting because you are taking on the feelings, burdens and emotions of other people. Remember it is actually a really great gift to have.

It is essential to figure out how to shield yourself from getting dragged down with negative energies and figure out how to practice self-care to discharge the negativity.

Empaths see the issues on the planet can't be understood through hate, rather, they should be addressed with love and understanding.

CHAPTER 9
EMPATHS AND ADDICTION: FROM ALCOHOL TO OVEREATING

Why are empaths so susceptible to alcohol, medicate, sex, nourishment, betting, shopping and different addictions?

Empaths can become overpowered and overstimulated because of their extreme sensitivity. At the point when they feel excessive strain, some empaths self-medicate. If they don't have any idea how to deal with this over-burdening, they numb themselves to close off their thoughts and feelings and to decrease compassion; however, not everyone is aware of this fact.

There can be a significant expense involved when it comes to adapting to your sensitivities through addictions. They wear out your body, mind and soul, making ailments gloomy, and create more tension as you attempt to deal with an over-invigorating world. The best-case scenario is that addictions give temporary alleviation from tactile over-burdening; however, in the long haul, they stop working and will exacerbate your sentiments of being overpowered.

Self-Evaluation and Getting Support

Though not all alcoholics or addicts ingest others' energy, a large segment of them do. Tragically, many empaths remain undiscovered and don't understand how overstimulation and high affectability fuels their addictive practices. It's vital to decide whether you're adapting to your sensitivities by taking part in addictions. How would you know? Ask yourself the following questions:

- Have I, at any point, figured, "Life would be so much better if I didn't drink or binge eat?"

- Have I, at any point, attempted to stop binge eating or using substances for a month? Can I do this for a couple of days, notwithstanding my best expectations?

- Am I self-medicating to ease social tension or to deal with the pressure I take on from the world?

If you presume you are using alcohol, medications, indulging in food or other addictive practices to deal with the tangible over-burden of being an empath, set aside some time to think about how you can adapt by assessing the accompanying statements.

I turn to substances or different addictions when:

- I'm overpowered by emotions (mine or another's).

- I'm in emotional pain and feel confused, anxious, or

discouraged.
- My feelings are harmed.
- I feel awkward in my skin.
- I can't sleep.
- I feel genuinely dangerous in a circumstance.
- I feel censured, accused or dismissed.
- I feel shy, anxious or don't fit in socially.
- I'm staying at home, and I need the certainty to go out in public.
- I'm worn out and need a jolt of energy.
- I feel depleted by energy vampires.
- I need to get away and close out the world.

Here's the way to interpret this self-assessment:
- If you choose yes to even one explanation that means that right now you have sensitivities.
- Noting two to five yes's shows that you are moderately dependent on the substance such as narcotics. You use this to help medicate your physical over burden. you decently depend on the dependence on self-sedate sentiments of tactile over-burden.
- Noting at least six yes's demonstrates you are, to a

great extent adapting to sympathy by taking part in addictive behavior.

Alternatives to Self-Medicating: Strategies and Solutions

Self-awareness is freeing. No disgrace. No blame. By monitoring your addictive tendencies, you're increasing your energy and learning to adapt with your empathy. At that point, you can manage it. Here are a few stages to help oversee sensory over-load.

To begin with, it's essential to recognize your addiction. Truly evaluate: How much do I drink or take different substances week by week? How frequently do I indulge in adapting to feeling overpowered? Do I go to different addictions (for example, sex, love, betting, shopping, computer games, the Internet or excessive work) to bring down my uneasiness level or to shut off my sensitivities? Be caring for yourself. Check whether you self-cure your emotions. Self-curing even once per week or once a month demonstrates that you may have an issue with addiction.

Second, it's essential to understand that nothing outwardly - no substance, individual, employment or measure of cash— can cause you to feel good about yourself and your sensitivities. Happiness is a good activity. You should figure out how to know, love and acknowledge yourself. It can be a deep-rooted process of revelation. The more you run from your sensitivities, the more awkward you will get. As the Buddha stated, "There is no outer refuge."

Thirdly, for a continuous plan to address your compulsions, you should consider entering psychotherapy or going to 12-step program for help. It's essential to discover healing ways to establish a stable relationship with yourself as well as with other people. At that point, as an empath, you won't be helpless before the excruciating sentiments of sensory over-burden, and you will have the option of focusing yourself on discovering a liberating feeling of balance in your life.

A Link between Empathetic People and Addiction

While numerous individuals portray themselves as being empathic, being an empath goes above and beyond. Having empathy ordinarily means that your heart goes out to somebody. This causes explicit feelings. Individuals who are empaths are exceptionally delicate individuals who assimilate the feelings, energy and worry of the people around them. They feel others' feelings or physical manifestations and might not be able to channel it. Regularly misdiagnosed as having social anxiety disorder, empaths' brains are wired to be hyperactive. This is the thing that makes them assimilate the energy of people around them.

There are some positive reactions to being an empath. For example, the capacity to frame further associations with others and more a vivid imagination. However, it presents various difficulties too. As empaths retain the energy of everyone around them, they can be inclined to feel more pressured and become easily overpowered, depleted and overstimulated. The pressure and anxiety they retain from others can leave them feeling depressed, anxious, emotionally burned out. It can also increase the risks of building up an addiction.

Traits of an Empath

Empaths are profoundly defenseless to addictions. They are easily overpowered by others' energy and so they search for outlets through substances or practices. This includes consuming food, shopping binges and substance mistreatment which are on the whole practices that can prompt compulsive habits that can inevitably transform into addictions.

Having the option to recognize yourself as an empath can assist you with better comprehension. It can prevent negative practices from being formed. Some common traits of an empath include:

- **Empaths are exceptionally sensitive**: Empaths are frequently informed that they are excessively delicate or that they have to "toughen up." These sorts of individuals are extraordinary audience members, nurturers, and are giving, yet they can definitely have their feelings hurt.

- **They will, in general, absorb others' emotions**: Empaths can retain others' feelings, both good and bad. They will, in general, feel these feelings in boundaries. When feeling anger or anxiety, they can easily become depleted or sick. At the point when they are surrounded by energy, they can bloom and prosper.

- **Empaths are typically introverted**: Because of how they retain other's energy, large groups can be overpowering and hard to manage. Empaths will, in general, favor littler groups or one-on-one cooperation. Regardless of whether they appreciate a social trip, they regularly limit the measure of time they spend in it to forestall burn out.

- **Alone time is critical**: Empaths need alone time to energize. Being around others can be depleting, making alone time significant. It stops an empath from feeling overpowered and gives them the space to clear their psyche.

- **Close connections can be difficult:** Empaths now and then battle seeing someone since they dread they may lose their character in it. They frequently require their connections to dispose of common ideas about how a relationship function. They want more space so they can avoid feeling overpowered.

- **Empaths can be targets**: Empaths affectability and energy can make them targets for other character types. Regularly alluded to as "energy vampires," their conceited nature makes them channel an empath and causing them to feel vacant or useless. Energy vampires are like narcissistic individuals or individuals who are dramatic.

COPING MECHANISM FOR EMPATHS

Empaths can become depleted and fed up with their environment. They may look to using substances. Don't let this occur. An empath must create techniques for securing their physical and psychological well-being. This can be troublesome now and again in light of an empath's craving to help other people. Yet, it is to their best advantage to abstain from becoming overwhelmed.

A few different ways an empath can secure themselves include:

Set limits: It can be hard to define limits when you feel obliged to help other people through their agony, yet it is important to show you are valuable. If somebody is depleting you, it is imperative to limit your connections and discussions with them. Figuring out how to say "no" is fundamental. You are not required to have a motivation behind why. "No" is a complete sentence. Saving your energy before you are depleted is fundamental in preventing exhaustion.

Physical separation: If a few people consistently cause you to feel depleted, limiting your contact with them will help. Physical separation can help forestall feeling overpowered. If this isn't an option envisioning you are separated can help too. This may incorporate pictures of isolating yourself from an individual's negative feelings. This permits your psyche to leave any bad feelings behind.

Take into consideration alone time: Empaths need alone time to revive. If you are continually immersed in others' encounters, you may wind up drained and incapable of recovering. Investing in energy alone can help you reconnect with your internal identity. It is critical to decreasing outer impacts and boosts from different sources. Pondering, submerging yourself in water, or working on breathing activities will all assist you in realigning yourself.

Go into nature: Empaths benefit from investing energy in nature. Go to a clean and open place. This will help you to feel like you are shedding the heaviness of others' difficulties. Encountering these sensations in nature can assist you with energizing and discovering a feeling of inward harmony.

CONCLUSION

Psychic powers arrive in a wide scope of structures. A few psychics feel things and others see them. Some people get messages as pictures. It is up to the person to interpret these messages. Most mystics share this endowment of psychic empathy. Feeling what someone else feels is an extraordinary method of interfacing with customers on a whole different level. Psychics with Psychic empathy can rapidly slip into your mind. You see it through your eyes and then you can give answers from a different view. Psychic empathy takes into consideration profound and legitimate connections to the world. They have the instinctive skill to see the universe and feel the silence drawn from them. This is how they improve their abilities.

Psychic Empathy goes farther in examining an individual and realizing how to help direct them towards a superior, progressively mindful and enlivened life. Sentiments can't reveal everything to you. Clairvoyant compassion can't tell you what is in your purse or what color pen you use. Some mystics have this power. Rather it is the increasingly emotional type of understanding that reaches out to other people. A mystic gifted at this will be an exceptionally strong person. They will be easy to talk to. Empathic people are healers. They can work through the emotional obstacles they are battling with. By and large, a great clairvoyant and an accomplished empathic will have therapeutic advantages.

www.ingramcontent.com/pod-product-compliance
Lightning Source LLC
Chambersburg PA
CBHW050005230526
45465CB00003BB/1261